ANIMALS AS FOOD

THE ANIMAL TURN

SERIES EDITOR
Linda Kalof

SERIES ADVISORY BOARD
Marc Bekoff, Juliet Clutton-Brock, Nigel Rothfels

ANIMALS AS FOOD

(Re)connecting Production, Processing, Consumption, and Impacts

Amy J. Fitzgerald

Michigan State University Press
East Lansing

♾ The paper used in this publication meets the minimum requirements of ANSI/NISO Z39.48-1992 (R 1997) (Permanence of Paper).

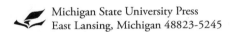 Michigan State University Press
East Lansing, Michigan 48823-5245

Printed and bound in the United States of America.

21 20 19 18 17 16 15   1 2 3 4 5 6 7 8 9 10

Library of Congress Control Number: 2014956942
ISBN: 978-1-61186-174-7 (cloth)
ISBN: 978-1-61186-175-4 (pbk.)
ISBN: 978-1-60917-462-0 (ebook: PDF)
ISBN: 978-1-62895-234-6 (ebook: ePub)
ISBN: 978-1-62896-234-5 (ebook: Kindle)

Book design by Scribe Inc. (www.scribenet.com)
Cover design by Erin Kirk New
Cover artwork is a detail from *Cracker Trail Cattle Drive* mural by Keith Goodson. Used courtesy of Lake Placid Mural Society, Lake Placid, Florida. All rights reserved.

g green press INITIATIVE Michigan State University Press is a member of the Green Press Initiative and is committed to developing and encouraging ecologically responsible publishing practices. For more information about the Green Press Initiative and the use of recycled paper in book publishing, please visit www.greenpressinitiative.org.

Visit Michigan State University Press at www.msupress.org

I would like to dedicate this book to my grandparents, Eileen and Eugene Schimmel, who taught me the value of family, hard work, and unconditional love.

Contents

Acknowledgments

My interest in the use of animals as food was piqued somewhat accidentally many years ago when I was in high school. I was wandering aimlessly through my school library, probably killing time between classes, when a thick, bright yellow hardcover book with stark red writing caught my eye. The title jumped out at me: *Animal Liberation*. I was sixteen years old at the time and had not been exposed to the term "animal liberation" before, much less the philosophy and politics behind it. I liked animals ("pet" animals really), so I checked the book out from the library, thinking that I would surely enjoy anything that had to do with animals.

Well, I can't say that I originally enjoyed the book. It troubled me. What Peter Singer wrote on those pages, along with the accompanying images, provided me with my first glimpse into how the food I was eating was produced. It challenged me to think more critically about the use of animals as food and even made me more attuned to what was going on in my community. I lived in a rural farming area where animals were raised to be killed for human consumption and crops were grown right behind my house to feed the animals in the industry. For the first time ever I started to wonder about what was going on in those large "livestock production" barns and how many animals were being raised in this way that required the production of all of that corn and those soybeans in my backyard for feed.

I eventually went off to university to study criminology and sociology, which at the time seemed rather far removed from my side interest in the use of animals as food. It was years later when I was a doctoral student determined to find a dissertation topic I was really interested in and that could sustain me through years of study that I was brought back to my interest in animal agriculture. I began to study the impacts of the changing animal agriculture industry, particularly the slaughtering sector, on communities and found that my sociological and criminological training helped me immensely in analyzing and understanding the changes in the industry and their impacts on human communities, the natural environment, and animals.

Years later, now as an associate professor, I teach sections about industrial animal agriculture in my Corporate and Governmental Crime, Green Criminology, and Current Issues in Criminology courses. I am sure that many students are surprised to find themselves addressing the topic in criminology courses, but it is certainly gratifying to see them making the connections in these courses as they really think about industrial animal agriculture, many for the first time. When I was asked to write this book, I was eager to do so, not only because of my interest in the subject matter, but also because I am keenly aware that the opportunities for consumers (such as my students) to gain their glimpses inside industrial animal agriculture are dwindling. Even someone raised in the midst of livestock and feed production, such as I was, can be surprisingly insulated from the practices surrounding them. I therefore wrote this book with two audiences in mind: those already interested in the topic looking for additional research and analysis, and individuals who have not yet had their first glimpse into industrial animal agriculture and the impacts thereof.

The research that has gone into this book is the culmination of many years of work, and there have been many people along the way that have helped to make this project possible, who I would like to take this opportunity to thank. First of all, I owe a debt of gratitude to authors before me who have written about various aspects of the use of animals as food: those who provided me with my first glimpses into industrial animal agriculture and those who have pushed my thinking further. I am also indebted to those who served on my PhD dissertation committee: Drs. Thomas Dietz, Linda Kalof, Aaron McCright, Angela Mertig, and Toby Ten Eyck. They challenged and guided me when I was making my first academic foray into this specific area of study. I would particularly like to thank Thomas Dietz and Linda Kalof, who provided me with invaluable academic opportunities and mentoring as a graduate student, and continue to do so today. I would further like to extend my appreciation to Linda Kalof for developing and editing the Animal Turn series with Michigan State University Press, and the staff at MSU Press for recognizing the value of the series and working to bring it to fruition.

I am also grateful to the University of Windsor for providing me with a sabbatical, which made it possible to focus on this project. The University of Windsor also provided the funding to hire an undergraduate student to help me compile the research for the book. I appreciate that funding and the hard work of Stephanie Oneschuk, who stepped in to fill that role. I would also like to thank the faculty and staff in the Department of Sociology, Anthropology, and Criminology for their support over the years, and the undergraduate and graduate students in the department, who through their interest and thoughtful questions have urged me to make the material and arguments I present in my classes on the use of animals as food accessible to a wider audience.

I am also indebted to Rochelle Stevenson, currently a sociology PhD candidate at the University of Windsor, who helped with securing permissions for the reproduction of the images found on the pages in this book, and who also helped me immensely with other projects so that I could focus my attention on completing this one. I would also like to collectively thank the individuals and organizations who granted permission to reproduce their images on these pages; I think these images help to illustrate and strengthen the arguments made herein.

Last, but certainly not least, I would like to thank my family and friends for their tireless support over the years. My friends, referred to collectively as the "Renegade Sociologists," have reminded me repeatedly over the years not to be confined by disciplinary boundaries, and certainly never to take myself too seriously. My family (i.e., immediate family, extended family, and family-by-marriage) has also provided me with invaluable support throughout the duration of this project and beyond. Knowing that I will always have their support means everything to me. In particular, I would like to thank my parents, Sandra Schimmel Fitzgerald and Reginald Fitzgerald, for encouraging me to pursue the route of higher education in the first place: it was a road that they had not traveled themselves, but that never stopped them from doing what they could to support me in my travels. I was also provided with companionship and (sometimes needed) distractions while working on this project by several canine and feline companions, who always seem to come into my life at just the right time and leave it even fuller: April, Bernie, Millie, Fawn, Zoe, Jasmine, and Emerson. Finally, I would like to thank my partner and husband, Sean Demers, for his emotional support and for being my strongest advocate.

Introduction

We seldom think much about the use of animals as food. On those rare occasions when we do stop to think about it, the scope and cultural significance of the practice gives it an aura of inevitability, which can stop a meaningful examination before it even begins. Before children have been thoroughly socialized into the "don't ask, don't tell" culture surrounding the use of animals as food, they may ask questions. Difficult questions. My sister-in-law recently told me that my four-year-old nephew had started inquiring about why people (including himself, his parents, and his brother) eat animals. I asked what her response had been. She said she had told my nephew that some animals are for food and that's just the way it is. This perception of inevitability in not unique to my family: it is cultural, and it militates against questioning it or trying to better understand it. It is the cultural equivalent of parents telling their children "because I said so" when they ask for the reasoning behind their parenting decisions.

The use of animals as food is all around us; in fact, it is our most common use of animals. The ubiquity of it, combined with the seeming inevitability of it, makes it an interesting if not challenging topic to write about. While writing this book I have been reminded of some of the difficulties I (and presumably others) encounter in teaching about gender: like the use of animals as food, gender is all around us and the way that gender is currently socially constructed is commonly assumed to be inevitable. In acknowledgment of the difficulties students have in studying something that is so taken for granted, when I teach gender content in my courses I begin with a quote by sociologist Judith Lorber, who remarks that "talking about gender for most people is the equivalent of fish talking about water."[1] I use this quote to illustrate for my students how difficult it is to think critically about gender because we take so much of its current manifestations as "natural" or as "fact." I also use it to prepare students for the discomfort they may experience: beginning the process of thinking about something that surrounds you, that you were born into and grew up in—like the fish in the water—can be challenging and even uncomfortable at times.

At this point you might be wondering what my nephew's inquisitiveness and the pedagogical exercise I use with my students have to do with the subject matter of this book. Well, I am asking you to be open to questioning the seemingly inevitable and to approach the material in this book as I ask my students to engage with the material in my classroom: with a willingness to critically examine that which surrounds you and mindfulness about when discomfort arises and why. There are few things that we take for granted as much as what we eat. For some of us, the food we purchase from the store requires such little thought that it feels almost instinctual. Sociologists have been arguing for years, however, that much of what we consider to be instinctual or natural is actually socially constructed.[2] Employing these insights, Carol Adams argues that just as "naturalness" has been used as a justification for the oppression of women, it is used to justify the use of animals as food: "Meat is a cultural construct made to seem natural and inevitable."[3] Understanding how the use of animals as food has been constructed and how it varies over time—both goals of this book—makes it easier to deconstruct the perceived naturalness and inevitability of the practice in general, and current production and processing methods more specifically. In short, it makes it possible to begin asking questions.

This book explores several questions, including the following: Why are animals used as food? How did the domestication and production of livestock animals emerge and why? How do current modes of raising and slaughtering animals for human consumption differ from earlier time periods? What are some of the consequences of the current mode of production? And what can be done to mitigate and ideally even reverse the impacts? The responses to these questions are numerous and complex. This book provides insight into the historical, cultural, political, legal, and economic processes that shape our use of animals as food, and in doing so contributes to answering these important questions.

THE SIGNIFICANCE OF THE USE OF ANIMALS AS FOOD

Before the questions articulated above can be addressed, a more immediate question undergirding the purpose of this book must be tackled; that is, Why bother examining the use of animals as food in the first place? I offer some preliminary responses here, which will be further developed throughout the book. First of all, residents of the United States, along with other western countries, have demonstrated a significant and growing interest in animals in general. The sheer volume of cute animal videos shared online is evidence of this interest; these videos and our growing cultural attentiveness to animals, however, are focused on companion animal species, such as cats and dogs. Yet the average person's most frequent contact with animals is with those that have already been killed and are served as food. Given the growing importance of at least some species of animals in western cultures, it makes sense to better understand our relationships with animals, particularly how some and not others came to be used as a food source, which has turned into our most common relationship with animals more generally.

The second reason why examining the use of animals as food is worthwhile is that an immensely profitable and powerful industry has developed around this use. The food industry is the second most lucrative industry in the United States (second only to the pharmaceutical industry).[4] U.S. residents spend one trillion dollars on food annually, a value that exceeds even that spent on automobiles.[5] Meat makes up a sizable portion of these expenditures, and the U.S. diet has become particularly meat heavy. Per capita meat consumption in the United States has increased dramatically over the years since colonists first brought livestock animals with them to the shores of North America. Only recently have per capita consumption rates nationally begun to inch downward; however, as the population in the United States and globally continues to grow, so does the gross amount of meat consumed. The number of animals used as food is so immense that it can be difficult to conceptualize. On an average day in the mid-1960s, eight billion "livestock" animals were alive globally, and ten billion were being slaughtered per year. Today, there are an astounding twenty billion livestock animals alive and fifty-six billion slaughtered per year. These already immense figures are expected to double by the year 2050.[6] An industry of this size, that controls what the majority of the population (at least in the United States) eats and is expanding its global reach, warrants examination.

Third, the use of animals as food has changed dramatically over time, and in order to understand current practices it is necessary to appreciate how and why the provision of animals as food has developed the way it has. For a significant amount of time, the consumption of animals by our ancestors took the form of scavenging. Scavenging gave way to hunting, as it became increasingly feasible, and then hunting for meat became less common as some species were domesticated and farmed as livestock. This originally took place as *subsistence livestock farming*, whereby families raised livestock as needed for their own consumption. In industrialized countries this mode of production became nearly entirely supplanted by *intensive livestock farming*. This mode of production is different from the former in that the animals came to be valued as commodities and farmers produce more than is needed for the consumption of their own family and sell the excess for profit. With the goal of maximizing those profits, producers (the term *farmer* becomes less appropriate as production is increasingly transferred from the hands of individual farmers to corporations) adopt measures to increase production and reduce costs, which gives rise to the current mode of production, *industrialized livestock farming*.[7] This current mode of production and processing can seem inevitable because most of us know no different. It is important to consider this larger historical trajectory in order to understand what this mode of production is and what it is not. Importantly, today "the goal of livestock production is to increase profitability, not produce for human needs—or, for exchange values, not use values . . . this is a highly peculiar way of producing animal-derived food compared to other socioeconomic periods."[8] Yet many are unaware of the unique characteristics of the current practice of using animals as food.

The fourth reason for examining the use of animals as food worth mentioning here is that, in spite of the cultural and corporate barriers erected to prevent people from thinking about the use of animals as food, there is evidence of growing cultural curiosity and unease. In his book *The Omnivore's Dilemma*, Michael Pollan notes that for centuries, our culture was able to assuage any discomfort felt about using animals as food, yet this seems to be changing. He writes, "Eating animals has become morally problematic, at least for people who take the trouble to think about it."[9] Pollan was certainly on to something when he wrote this nearly ten years ago. There is a growing critical consciousness about where our food comes from. His book, which addresses the central question of what we should eat, reached a wide audience and was one of the *New York Times*'s top ten books of the year. It seems that increasing numbers of people want more information about how their food is produced.

The fifth reason why an examination of the use of animals as food, such as that rendered in this book, is needed is that the industry responsible for providing the vast majority of the animal-derived food products has not been forthcoming about the questions consumers want answered. Consumers in the United States in particular receive insufficient information about meat production and processing. This is evidenced by a study examining a nationally representative sample of residents that found that the majority reported not having a source of information for the welfare of animals produced for food.[10] In contrast, a study in western Europe found that approximately two-thirds of respondents believed they were well informed about animal welfare.[11] Why is it that few people in the United States are privy to how the animals (and the by-products) they eat are raised? The industry in the United States is very powerful, and it has a vested interest in maintaining a healthy distance between their products and consumers. The industry has even leveraged its power to get laws passed in some states criminalizing taking photos or recordings within animal production and processing facilities. Commentators have decried these actions. For instance, Pollan argues that consumers have a right to see inside farms and slaughterhouses, and an op-ed piece penned by a law professor and recently published in the *New York Times* titled "Open the Slaughterhouses" demands the curtain be lifted, yet this seems unlikely to happen anytime soon.[12] Until such a time that consumers can create their own firsthand accounts, they have to rely on secondhand accounts, such as this book.

In fairness to the industry though, many consumers are not interested in learning more about what is underneath the cellophane-wrapped meat they purchase from the grocery store and the process that transformed it from a living animal to a piece of meat. As French anthropologist Noëlie Vialles succinctly explains, "we demand an ellipsis between animal and meat"; we do not want to confront meat as the animal(s) it once was.[13] In one of the classes I teach I indicate on the course outline when we will be watching specific films. I have noticed that attendance is lower on the day we are scheduled to watch a film about the production of meat. Information of the sort contained in such films undercuts the important ellipsis that Vialles refers to and may be avoided as a result.

There is a similar reluctance to confront the consequences of the use of animals as food, yet these consequences are becoming increasingly difficult to ignore. I would therefore suggest these pressing consequences as the final reason why the use of animals as food is worthy of examination. The consequences include growing animal welfare challenges, negative environmental impacts, risks to human health, and negative impacts on workers in the industry and in communities more generally. These impacts usually receive scant attention because, as Weis explains, "the deceptive efficiency of industrial capitalist agriculture and its manifestation in cheap, bountiful food have long overshadowed the instability and inequalities of the system."[14] This book aims to pull back the curtain (which in this case takes the form of barbed wire, electrified fences, sophisticated locks, and security systems; see figure 1) that shields the industry and protects the status quo in order to better assess and understand the impacts, including how they have changed over time.

This is by no means an exhaustive list of the reasons why examining the use of animals as food in a context such as this is worthwhile; these are simply the six main reasons that spurred me to write this book. My interest in the well-being of animals has for years drawn me to the topic of how they are used as food, because this is the purpose for which most domesticated animals meet their end. That being said, as will be demonstrated throughout this book, animal agriculture ought not to be of interest only to those concerned about the welfare of animals and the ethics of consuming them. As Singer and Mason point out, no other human activity impacts the earth more than animal agriculture: it affects every living being on the planet, at least indirectly, as well as the planet itself.[15] The topic therefore rightfully resonates with many environmentalists. I would suggest that the state of animal agriculture today should also be of interest to those concerned with human health, workplace health and safety, immigrant rights, and general community well-being. I make the case for the coalition of these interests throughout the book.

Concern about animal agriculture and the associated issues listed above are commonly considered to be the province of those closer to the left end of the political spectrum; however, there are reasons why those who are more conservatively inclined ought to be concerned as

Figure 1. Feedlot in Wauchula, Florida, protected by several layers of fences. (Source: ©2013 Amy Fitzgerald)

well. For instance, Scully suggests that conservatives should be interested in changing the face of animal agriculture because of their longstanding concern about matters related to morality, tradition, and the value of life.[16] I would add that fiscal conservatives ought also to be concerned about animal agriculture because of the amount of government subsidies currently funneled to the industry.

In addition to weaving together a wide variety of social causes and political perspectives, the information provided in this book also provides insight into more than just animal agriculture. It provides a valuable context and vehicle for understanding the broader social distribution of resources.[17] It also provides an entry point for examining significant historical processes, such as domestication, colonialism, industrialism, capitalism, and neoliberalism. In theory then, the state of animal agriculture should be of interest to a great many people, yet very few have made these connections, and many are downright resistant to doing so.

PRODUCTION, PROCESSING, CONSUMPTION, AND IMPACTS

The connections of the type detailed above can be made more explicit by bridging the distance in animal agriculture between production, processing, consumption, and their impacts. These are links in the supply chain of serving up animals as food that have literally and symbolically been torn apart. This rupture is a relatively recent problem. As Mench et al. explain, "There was a time in American history when this chain was very short indeed, extending perhaps 40 or 50 feet from the barnyard to the kitchen."[18] Today, this chain has been stretched to the point that it is difficult when focusing on a specific link in it to appreciate the entire chain.

The Scope of the Book

Previous academic books that have addressed the use of animals as food have tended to focus on one species (e.g., Rifkin focuses on cattle), or break the chapters down by type of animal (e.g., Horowitz), or focus on one aspect of the industry (e.g., Stull and Broadway focus on processing). These works have made significant contributions to the literature, and I draw on their insights with gratitude here. The aim of this book, however, is different. The purpose of this book is to provide a holistic view of the use of animals as food, which necessitates focusing on more than one species and more than one sector of the industry.

This book is also somewhat different in that it pays significant attention to the impacts of using animals as food in the current mode of production/processing/consumption and makes suggestions for redressing these impacts. Doing so risks crossing the line from description to prescription, something that academics are often wary of, and in the literature on animal agriculture it is ground that has been occupied more by popular writers, such as Foer and Pollan.[19] At the outset, I knew that I wanted this book to draw heavily on interdisciplinary research from fields such as sociology, (critical) animal studies, history, economics, law, political science, anthropology, criminology, environmental science, geography, philosophy, and animal science. After compiling research from these fields I felt compelled to include chapters on the impacts of industrial animal agriculture and recommendations for mitigating these impacts because it became increasingly apparent to me that as the use of animals as food and the industrialization of provisioning this food increases so do the impacts, and the evidence indicates that the problems are worsening. In short, something needs to change, and quickly; I felt that leaving that material out would be telling only half of the story of our use of animals as food.

That being said, due to space constraints, I could not cover everything related to animal agriculture. This book covers quite a bit of historical and topical terrain, and I have done my best to strike a reasonable balance between generality and specificity. In tackling a topic as immense as the use of animals as food, it is impossible to attend to all of the nuances, and some degree of generalization is inevitable.[20] So, while general trends are discussed herein, I am aware that there are always exceptions to the rule. I have also sought to narrow the focus as much as possible without losing sight of the larger picture. More specifically, the book focuses on the U.S. context, for a number of reasons. First of all, it has been a pioneer in the industrialization and corporate consolidation of animal agriculture and as a result is currently grappling with a number of consequences of these processes that I wanted to capture. Second, the United States is among the largest producers and consumers of animals as food. In fact, in 2003, it earned the title of the first country to produce over ten billion livestock animals in a year.[21] It therefore makes sense to focus on developments here. Although the primary focus is on the United States, there are portions of the book that necessarily draw on research from other countries, address the global significance of industrial animal agriculture, and trace historical developments elsewhere to the shores of the United States.

In the interests of further narrowing the material covered in the book, the focus herein is primarily on three of the most commonly produced and consumed animals in the United States: chickens, pigs, and cattle. Other animals, such as horses, fish, ducks, geese, and turkeys, are mentioned in a few places in discussing the ways that industrial animal agriculture is expanding (e.g., the increasing industrial production of fish, industrial processing of horses), but because the industrialization of animal agriculture and the consumption of animals as food has primarily taken place on the backs of chickens, pigs, and cattle, that is the primary focus here.

The Structure of the Book

A primary goal of this book is to provide a holistic picture of the use of animals as food and to explicate the connections in the supply chain that are obscured in the current mode of food production. In attempting to do so, however, it became apparent that addressing all of the links in the supply chain, consumption, and impacts simultaneously could be difficult and confusing for the reader. Therefore, I have divided the book up into chapters that focus on production, processing, consumption, and impacts respectively, while also addressing the connections between them.

Chapter 1 provides the historical information necessary to understand the development of animal agriculture. It ends at the point of industrialization, which is where the subsequent chapters pick up. Chapter 2 examines the industrialization of the production of animals as food and provides insight into just how much and how quickly the use of animals as food changed over a relatively short span of time. The third chapter provides the same type of analysis but focuses on the processing sector of the industry, or how the animals who are produced for consumption are processed into consumables. Chapter 4 explores the consumption of the end product: meat. It details how consumption levels and practices have changed over time and the relationship between production, processing, and consumption. Chapter 5 furthers the examination of the connections between the supply chain and consumption by detailing the interconnected impacts. The boundaries of this chapter were the most difficult to delimit, as the impacts of the industrial production, processing, and consumption of meat are wide-ranging. The chapter is therefore delimited by focusing on the impacts that have come to light and been identified as the most pressing. Building upon the holistic picture of the use of animals as food developed throughout the book, the final chapter addresses ways of mitigating and, ideally, eliminating the negative impacts and speculates on the future of industrial animal agriculture.

A few words about the use of language in these chapters are warranted. First of all, the term *animal* is used herein, including in the title, to refer to nonhuman animals. I acknowledge that the term *animal* is not ideal: it can obscure the fact that humans are animals and can contribute to a constructed distancing between humans and other animals. I employ the term here, however, because it is by far the most commonly used term in the areas of literature drawn upon in this book and because the proposed alternatives are rather cumbersome. Where possible, I refer to specific species of animals (e.g., pigs) rather than the encompassing word *animals* to mitigate at least some of the problems, although I do not claim this as a solution.

The term *livestock* is also used in some instances to refer to a group of animals from certain species produced for human consumption. This term is problematic because it reifies the property status of animals. Nonetheless, it does accurately reflect the way that animals are treated in animal agriculture, so I have not sought to sanitize it, although I certainly think it fair to problematize it.

The term *meat* is also used herein as it is the most common term used to refer to the parts of animal bodies deemed consumable by humans. The term is also used broadly here and is intended to include *poultry*, which is not always the case elsewhere.[22] I acknowledge that this term can obscure the reality of the specific body parts that are being consumed. Some have instead suggested the use of more specific, alternative terms, such as *remains, flesh,* and *corpse portion*.[23] Although I do not use these more evocative terms, where possible I do reference the specific type of animal and body parts being consumed to remind the reader where the consumed item derives from. Although I acknowledge that the term *meat,* like *animal* and *livestock,* is problematic, I employ it in this book in an attempt to avoid confusing, and even alienating, the readership. My hope is that by the end of the book the reader will better understand why these terms are problematic, which would not be possible if I lose them at the outset.

It is also useful to provide a definition of *agriculture,* which I have found to be infrequently provided in literature on the subject. Although it is often assumed to refer strictly to materials produced for food, that is not necessarily the case; it can include materials such as cotton and tobacco.[24] Additionally, the term covers both plant and animal materials produced for food.[25] In one of the more comprehensive yet concise definitions, Woods and Woods define agriculture as "the production of plants and animals useful to humans."[26] This book is focused primarily on animal agriculture, although it should be noted that the current mode of production could not exist without the production of plants as feed crops.

The term *industrial animal agriculture* is also used in this book to identify a more specific and recent form of animal agriculture. The origins of this mode of agriculture and its characteristics will be discussed at length herein. The Food and Agriculture Organization's definition provides a quick entrée into what it entails.[27] Their categorization of industrial animal agriculture includes those systems that purchase at least 90 percent of their feed, meaning that the animals are not grazing or foraging: the food is being brought to them. This often coincides with intensive production, where animals are kept in confinement.

Confronting the Use of Animals as Food

Philosopher Peter Singer and attorney Jim Mason point to transparency in animal agriculture as an essential part of ethical consumption. They argue that consumers should have access to information about how their food is produced and processed, and I also suggest that they ought to have access to information about the impacts thereof. The other ethical principles they recommend vis-à-vis animal agriculture include that it be produced fairly, meaning that the costs of producing meat and other animal products should not be borne unfairly by some individuals in the form of greater exposure to the externalities produced by the industry (such as negative environmental impacts); those who work in the industry should not be made to suffer as a result; and animals used by the industry should not be made to suffer for minor reasons. In short, they recommend that people only consume animal products whose production they can witness in person.[28] Perhaps most people would agree that these requirements are reasonable. Yet in practice this has become much more difficult over time.

As will be detailed throughout this book, changes over time in the connectedness between the production of livestock animals, killing and processing the animals into meat, and consumption have created fundamental challenges to Singer and Mason's requirements for ethical consumption. I suggest you keep Singer and Mason's recommendations for ethical consumption in mind as you read through the subsequent chapters. The ability to approximate their recommendations provides a sort of measure and reminder of how much the use of animals as food has changed over time. I also invite you to consider the ways suggested herein for (re)connecting production, processing, consumption, and impacts in the practice of using animals as food; doing so provides a route around the road blocks posed by the ubiquity and assumed inevitability of it. To return to Lorber's analogy, seeing the connections between the constitutive elements in the use of animals as food will assist us in seeing the system that surrounds us, just as visualizing points of connectivity would make it easier for the fish to better focus on the water that surrounds him/her.

Prehistory through the Colonization of North America

Did our ancestors consume large amounts of meat? How did the domestication of some species of animals unfold? When did agriculture develop and why? These are some of the questions this chapter addresses. In doing so, it traverses a large span of historical terrain. The chapter examines the relationship between humans and the animals they consume, all the way from our hunting and gathering ancestors through the introduction of domestication and into the development of subsistence animal agriculture. It explores the use of livestock animals not only as sources of food but also as sacred symbols and tools of colonization. Collectively this material provides the historical grounding necessary for the subsequent chapters that explore the more historically proximate production of animals used as food, the processing of these animals into meat, and the consumption of the end products. In order to fully understand the industrial animal agriculture of today it is necessary to first understand how we arrived at this point in history.

HUMAN PREHISTORY

Certainties about what transpired in early human history are hard to come by due to limited evidence. Based on the evidence that is available, we can draw some tentative conclusions about meat consumption and human relationships with animals. Our earliest direct ancestor, the australopithecine (living from one to 4.2 million years ago), apparently consumed very little, if any, meat.[1] This finding runs counter to the common assumption that our early ancestors consumed significant amounts of meat. In particular, it was believed that the first individuals in our genus, *Homo*, were avid hunters who consumed large amounts of meat. This belief was based on archeological findings of early tools with the remains of large animals from that period, which was interpreted as evidence of hunting. Further examination, however, indicated that the animal bones had been gnawed on by nonhuman animals prior to being cut by the stone tools. The conclusion drawn from this more recent discovery is that early ancestors in our genus were not actually hunting but were scavenging and using tools to do so.[2] Although based on this evidence we cannot entirely rule out the possibility that these ancestors did hunt, it is safe to conclude that "the earliest members of our genus were not great hunters of wild beasts, but largely sneaky scavengers."[3]

The earliest unambiguous evidence of humans hunting to procure meat dates back to between four hundred thousand and five hundred thousand years ago. Space constraints prohibit a detailed discussion of this subsistence hunting period. It is important to note, however, that our ancestors were likely still scavenging for at least some of this time as well.[4] The animal remains found at *Homo* community sites provide evidence in support of the assertion that our ancestors were still scavenging at this point in time: bones of several different types, ages, and sizes of animals were found in these communities, whereas the remains in caves of predator animals (e.g., lions) are more uniform and come from the young and weak animals who are easier to hunt.[5] The archeological evidence from much of human prehistory therefore points to something quite contrary to the commonly held view that humans have always subsisted on meat and that our ancestors were virulent hunters. Instead the evidence suggests that humans eventually integrated meat into their diets "despite a strongly herbivorous ancestry," not the reverse.[6] The transition from scavenging to hunting certainly precipitated significant changes in the diet of humans and the relationship between humans and animals. However, many more dramatic changes in human-animal relations and human consumptive practices were yet to come. A particularly important shift, and one that is central to the topic of this book, took place hundreds of thousands years later: the move from hunting animals to domesticating them.

DOMESTICATION AND AGRICULTURE

Before delving into a description of how humans domesticated animals and why, it is worth providing a definition of domestication itself, as the term is frequently utilized in an overly restrictive manner. Anthropologist Barbara Noske provides us with an encompassing understanding of domestication as "that situation where humans force changes on the animal's seasonal subsistence cycle."[7] I employ this definition herein because it includes relatively minor and major human intrusions in the lives of animals and makes it possible to appreciate the nuances involved in the process of domestication. This process is very gradual, and as Noske points out, in prehistoric times the distinction between wild and domestic animals was not as dramatic as it is today.[8]

The specifics of when and why the process of animal domestication began are still being debated. Most agree that the process was underway approximately ten thousand to twelve thousand years ago during the Neolithic period and that the first animal to be domesticated was likely the wolf, from whom the domestic dog descends.[9] However, there is also some speculation that the horse may have been domesticated earlier, perhaps during the Upper Paleolithic.[10] Evidence in support of this hypothesis includes the location of horse teeth from this time period that display evidence of having been worn down, presumably from continually chewing on something that was restricting movement, as well as a carving from this period at St. Michel d'Arudy of a horse with lines on his/her face that have the appearance of a harness.[11] While the debate over which animal was domesticated first is interesting, it is predicated on the assumption that domestication is an either/or proposition that abruptly occurs: thus, there can be first- and second-place finishers. Instead of conceptualizing domestication as a discrete event that occurs at a specific point in time, it is more helpful to conceptualize it as a process, consistent with the definition of domestication articulated above, where the degree of domestication occurs along a continuum.[12]

In addition to dogs and horses, the earliest animals to make their way toward the totally domesticated end of the domestication continuum include goats, sheep, pigs, cattle, and chickens. There is solid evidence of the domestication of goats nine thousand years ago, sheep shortly thereafter, pigs six thousand to eight thousand years ago, and cattle around 6500 BCE.[13] These animals were conducive to domestication because, to varying degrees, they live in groups, can recognize a leader among their group, and have relatively slow responses to potential danger.[14]

The significance of the domestication of these species cannot be overstated. According to Juliet Clutton-Brock, an authority on animal domestication, the shift from humans hunting animals to domesticating them "may be seen as the most important change in social and cultural behavior to have occurred throughout the history of the human species."[15] One of the main reasons why this shift is so significant is that for the first time people had the power to control their own food supply. Although they were still vulnerable to environmental fluctuations, such as climate changes and pest invasions, the increased control they had over their food supply mitigated some of the impact of these fluctuations.[16] The domestication of animals also made it possible to eventually use them for traction. Evidence indicates that animals were first used to pull ploughs in 4500 BCE.[17] This technological development made plant-based agriculture more practicable.

Domestication may have also had negative consequences for people. David Nibert conceptualizes domestication as resulting in the oppression of animals and employs the term *domesecration* to refer to "the systematic practice of violence in which social animals are enslaved and biologically manipulated, resulting in their objectification, subordination, and oppression."[18] Nibert argues that animal domestication also facilitated the oppression of human groups. He notes that Roman armies were able to be successful in their military endeavors because they brought their cattle with them on military expeditions so that they had a ready supply of food. Prisoners of war were then made to work on cattle ranches. The domestication of animals and human warfare may therefore have been mutually reinforcing.

Although there is some debate regarding the ultimate consequences of domestication, there is general agreement that agriculture is actually more labor intensive than hunting, scavenging, and gathering. Academics have therefore sought to understand why our ancestors undertook these more labor-intensive activities. Numerous reasons for the transition to agriculture have been theorized. To date there is general consensus that three primary factors were involved: climate change, population pressures, and changes in the organization of human societies at the time.

The first evidence of agriculture dates to after the last ice age, when temperatures rose. It is believed that this change in climate made the environment more hospitable to agriculture. During this time there was also an increase in the human population that would have driven up the number of animals being hunted, making hunting more competitive and reducing the populations of hunted animals. This population pressure could have also contributed to the transition to agriculture, which is conducive to higher population densities. There is further evidence of significant changes and growing complexities in human societies at this time. Socially complex groups of hunter-gatherers appear to have transitioned to agriculture before less complex groups. The more complex groups were also more sedentary, which may have made agriculture more attractive to them. Concomitantly, there was a shift in economic organization from the level of the community to the household, which may have facilitated a shift from communal hunting and community-wide sharing of the proceeds to a focus on immediate familial or individual well-being, which could be fulfilled through small-scale agriculture.[19]

Although the exact roles that these factors played in the transition from hunting and gathering to agriculture cannot be articulated definitively, it can be unequivocally stated that the transition forever changed human societies and the relationship between humans and animals. For better or worse, it brought humans and animals into closer contact and made each more dependent on the other materially, and in some ways spiritually.

WORSHIPPING AND COMMODIFYING ANIMALS

As our ancestors were in the process of domesticating certain species of animals some groups of people were engaged in worshipping them. Cattle in particular played an important role in some spiritual belief systems.[20] The fact that cattle are the most commonly depicted image in the earliest recovered human drawings, such as in the Lascaux caves in France, points to their spiritual and symbolic importance. Ancient cattle shrines have even been discovered, such as one in present-day Turkey that dates back to approximately 6000 BCE. Linda Kalof asserts that cattle were so culturally important that the bull was the most significant representation in art of the third millennium through most of the world. The bull was revered as a symbol and even facilitator of fertility. This cultural significance did not mean that cattle were protected from harm. Cattle were simultaneously revered and killed: they marched in religious processions and were sacrificed in religious rituals. Rifkin refers to this historic period of worshipping cattle as "the cult of the bovine."[21] These bovine cults extended from Egypt to east and south Africa and could also be found in the pre-Christian era in present-day Jordan, Syria, Lebanon, Israel, Greece, and Italy.[22]

In repudiation of the bovine cults, Christianity modified some of the rituals and framed the bull god as a symbol of evil. In 447 CE, the church gave a first official description of the devil as "a large black monstrous apparition with horns on his head, cloven hoofs—or one cloven hoof—ass's ears, hair, claws, fiery eyes, terrible teeth, and immense phallus, and a sulphurous smell."[23] This description certainly evokes the image of something at least similar to the bull.

Although being sacrificed in religious rituals and depicted by the Christian church as the devil were not positive for the well-being of cattle, it is the later association with economic value that I argue would become the most dangerous for cattle and their kin. Rifkin points to the Kurgan warriors as a driving force in the shift from revering cattle to commodifying them. The Kurgan people inhabited the Eurasian steppes (a region surrounded by present-day eastern Europe, the Ukraine, Mongolia, and Manchuria) and are known for being the first group to breed horses to carry people. Capitalizing on this technological development, they were able to begin invading Europe, India, Iran, and Scandinavia around 4400 BCE. They invaded to secure grazing land for their cattle and horses, and while they were there they expropriated cattle from local populations. This took place over a period of three thousand years. In the process of expropriating cattle, the Kurgans commodified and morphed them into a form of transferable capital. This commodification contributed to a significant rift between the warrior and priestly classes because among the Kurgan people the warrior class had come to attribute an exchange value to cattle, whereas the priestly class still saw them as sacred.[24] The warrior class prevailed, and over the span of a few thousand years cattle widely went from being divine to being a commodity. The eventual fixture of this status is illustrated by the fact that the word *capital* originated in reference to a head of cattle.[25]

This is an infrequently told history, yet it is significant. Although few people today are likely aware of the invasions and expropriations undertaken by the Kurgan people thousands of years ago, the impacts are still felt today. They turned cattle into a transferable form of capital, and in doing so "helped prepare the economic ground for modern capitalism and the colonial era in world history."[26] Rifkin goes even further than asserting that they laid the foundation for modern capitalism and argues that these Eurasian herdsmen were actually the first protocapitalists.

Rifkin is not alone in suggesting that the origin of capitalism is rooted in the keeping of livestock and farming more generally. Others have suggested a connection between the development of agriculture, capitalism, and social inequality.[27] In her book *The Origin of Capitalism*, Ellen Wood asserts that the development of agriculture laid the groundwork for the division of labor and the ensuing dependence on others for the provision of food.[28] Tasks became more specialized, and people no longer produced their own food. As this process unfolded, property ownership became concentrated in the hands of fewer and fewer people, and many of the propertyless, often referred to as tenant farmers, turned to working the land for property owners, producing food in exchange for a wage. The property owners then sold the food on the market, from which the workers used their wage to purchase their sustenance.

According to Wood, capitalism emerged in the midst of this process because food production became governed by market imperatives instead of subsistence needs. This transition to producing based on market demands was augmented by urbanization, as fewer people lived on the land and produced their own food, and therefore purchased it from the market. As evidence of her assertion that capitalism was born out of the development of agriculture, Wood contrasts the development of the Dutch Republic and England. The Dutch Republic was positioned to be the first nation to make the transition to capitalism; however, it was stymied by national disinvestment in agriculture during the seventeenth century. In contrast, England continued to invest in agriculture, and Wood argues that, as a result, the nation surpassed the Dutch Republic and began the transition to capitalism.[29] Not only did landownership become increasingly concentrated in England, but productivity among tenant farmers increased because those who produced more had better access to land and paid more favorable rents. Less productive tenant farmers lost access to property and had to sell their labor. The result was that agriculture in England became increasingly productive, and it became possible to feed a large population that was not directly involved in food production. This freed people up to work for a wage elsewhere and to become consumers. Wood's focus on the countryside as the birthplace of capitalism is somewhat unique.[30] There is not agreement among historians regarding the birth of capitalism; however, much of the focus in the literature has been on urban areas, and many argue that merchants and manufacturers were the driving force behind the transition to capitalism.

Regardless of exactly how and where the development of the capitalist form of property relations unfolded, it brought with it profound changes, including to the relationship between humans and animals (particularly those held as livestock). While animals were certainly exploited under other modes of production, the degree of exploitation became almost unfathomable under capitalism, and it has been suggested that the exploitation of animals and people under capitalism are inextricably connected.[31] Longo and Malone assert that the alienation of humans from nature under capitalism facilitated the treatment of animals as commodities. In short, the distancing of humans from nature made the subjugation and commodification of nature and its animal members by people and institutions more acceptable.

The term *alienation* is frequently used to refer to the condition of human workers under capitalism, not just in relation to nature. The concept has also been applied to animals under capitalism. Noske argues that the alienation animals experience under capitalism manifests itself through deanimalization, whereby animals become virtually unrecognizable; they are "mere body parts akin to machine parts in the juggernaut of capital."[32] This deanimalization is discussed in detail later in the book in the context of describing the living conditions of animals used by industrial animal agriculture, wherein their natural animal behaviors are thwarted. In order to understand this process it is necessary to take a few steps back and explore the changes that have occurred vis-à-vis the commodification of agriculture more generally under capitalism.

AGRICULTURE UNDER CAPITALISM: THE REDUCTION OF FOOD TO COMMODITY

There has been a great deal of discussion in the literature regarding how to conceptually parse the historic changes in agriculture over time. Some speak of an agricultural revolution.[33] Others argue that the term *revolution* implies that there has been a significant departure from the practices of the past and suggest that the term not be used because there is continuity between the economic logic guiding agriculture today and that in the centuries prior.[34] Still others have endeavored to break the transition to modern agriculture up into phases. Such heuristic devices can be useful for gaining an appreciation of the changes that have occurred over time, as long as one avoids interpreting the delineated phases or revolutions as discrete or mutually exclusive. With this qualification in mind, F. M. L. Thompson's work provides a useful mapping of the changes in agriculture.[35] He identifies four revolutions or phases characterized by technological and economic changes, which provide insight into the changes animal agriculture underwent over time.

He identifies the shift from subsistence farming to farming for the market, guided by market demands, as the first phase. Recall that Wood points to this as the origin of capitalism. Animal agriculture became prolific during this period.[36] Because people were moving to urban areas and plagues were reducing the human population, it became increasingly difficult to find people to work farming plant-based foods. Landowners therefore embraced production that required less labor power: raising and killing livestock animals for food.[37] An increasing amount of land was consequently dedicated to raising livestock animals. The proliferation of animal agriculture was also tied to broader economic conditions at the time. In his classic book on the origins of capitalism, Wallerstein states that during this period England and Spain increased the amount of land they devoted to pasture in part because they noted that the price of wool and meat had been more resilient than crops were during the depression in the fifteenth century.[38] The increase in sheep farming then precipitated the great enclosure movement in the sixteenth century, which furthered the concentration of property ownership.

The second phase, which Thompson refers to as "the agricultural revolution proper," included the enclosure movement, the introduction of crop rotation, and attempts to improve livestock animals. It coincided with an increase of capital and labor brought to bear upon the land. He suggests this revolution was completed in Britain by 1815.[39] The third phase immediately followed the second and spanned from 1815 to 1880. During this period farming moved from being a simply extractive industry to taking on characteristics of the manufacturing industry. Fertilizers and artificial animal feed were introduced. During this period value was added to raw materials through processing, and farming became commercialized.

Thompson suggests there were no significant changes during the period between 1880 and 1914 and asserts that the fourth phase began in 1914. This phase is characterized by mechanization and the increasing use of fertilizers. He suggests that a potential subdivision of this phase, or perhaps a new, fifth phase, is marked by the application of intensive methods to livestock farming.[40]

Although Thompson's categorization and the like are useful for providing a sense of how agriculture has been transformed across these time periods, and forms a basis for understanding the developments discussed in the subsequent chapters, the attention paid to animals is restricted to their roles in agriculture; their related role in colonization is commonly overlooked. Attending to the role of animals in colonialism is particularly important for understanding the development of animal agriculture in the territory that was to become known as the United States.[41]

ANIMALS AND COLONIALISM

The history of capitalism is one of expansion. Ellen Wood eloquently explains it this way:

> These imperatives [competition, accumulation, and profit maximization], in turn, mean that capitalism can and must constantly expand in ways and degrees unlike any other social form. It can and must constantly accumulate, constantly search out new markets, constantly impose its imperatives on new territories and new spheres of life, on all human beings and the natural environment.[42]

Part of this expansion meant travelling to new lands, and agricultural animals were part of those colonial expeditions. By the sixteenth century, overgrazing became a problem in western Europe.[43] As a result, the British turned to Scotland and Ireland to graze animals. When they required more land, they turned their efforts to North America, Argentina, New Zealand, and Australia.[44]

The colonists brought livestock animals with them to the plentiful lands of North America. However, the movement of animals between western Europe and the North American colonies was not unidirectional: colonizers brought some animals from the colonies back to Europe with them. For instance, in the sixteenth century European colonizers brought domesticated turkeys back with them from the Americas.[45] The movement of animals, however, was primarily in the other direction, from Europe to the colonies. For instance, the Spanish brought longhorn cattle with them to the Americas. Due to the favorable conditions, the cattle rapidly reproduced, and the population increased to the point that the Spanish needed help to control the cattle. They therefore taught members of the native populations how to ride horses so that they could assist them with the cattle.[46] Other species of animals brought to the new world included horses, dogs, donkeys, sheep, goats, and chickens. Due to the favorable conditions, their numbers also grew rapidly.[47] They populated the land so quickly and thoroughly that Virginia Anderson remarks, "in a real sense these creatures, even more than the colonists who brought them, won the race to claim America as their own."[48]

Before long the increasing human and livestock populations in the North American colonies placed too much pressure on the land. The growing demand for meat domestically and internationally combined with overgrazing contributed to significant changes in the ways food animals were produced beginning in the second half of the nineteenth century. Subsequent sections of this book detail these changes by building upon the foundation laid in this chapter regarding meat consumption by our ancestors and how some species of animals came to be relied upon as sources of meat via the processes of domestication and agriculture and as sources of profit via capitalism.

Putting the subsequent chapters into context requires keeping our long history with animals in mind, only part of which entailed interacting with animals as food:

> Three million years ago animals were just predators of our australopithecine ancestors. By two million years ago, the first members of our genus were scavenging on animal tissues; by 500,000 years ago the large brained early humans were hunting animals. But it was only in the last 100,000 years, and possibly just the last 30,000, that animals came to occupy that immensely diverse set of roles in our society that they have today: animals as sources of food; animals as companions; animals as the subject of stories, myths and paintings; animals as metaphors; animals as objects to abuse.[49]

The reliance of humans on animals as sources of food is therefore only fairly recent in human history, yet from the current historical vantage point it is difficult to imagine it any other way. Further, as detailed in the subsequent chapters, the current industrial form of producing meat once put into perspective has an extremely brief, albeit interesting, history. The overwhelming dominance of this mode of production today makes it difficult to fathom any alternatives, and it can even contribute to a feeling of apathy. Cultivating an understanding of the historical development of the use of animals as (industrial) food—a process begun in this chapter and carried through the subsequent ones—contributes to destabilizing the cultural scripts of inevitability, normality, and progress.

The Industrialization of Livestock Production

In the third phase of F. M. L. Thompson's chronology of agricultural developments, a significant shift took place between approximately 1815 and 1880: agriculture adopted the characteristics of the manufacturing industry and became commercialized.[1] This shift would forever change the way livestock animals are raised. In fact, at this point in time it begins to make more sense to refer to it as the way livestock animals are *produced* rather than to use the term *raised* because traditional husbandry practices gave way to an industrial logic, and industrial animal agriculture was born. Deborah Fitzgerald, a professor in the Science, Technology and Society program at MIT, points out that the general public may think the application of the term *industrial* to describe current food production peculiar. Yet this is exactly what it became at this point in time. Although the term *industrial* is appropriate in this context, it does tend to get thrown about with little reference to the precise features of animal agriculture that warrant its use.

For clarification we can turn to Hardeman and Jochemsen who delineate five main characteristics of industrial agriculture.[2] The first is *mechanization*, where human labor is replaced by machines. Under industrialized animal agriculture it is no longer necessary for humans to feed and water their livestock animals—it is automated, which renders this former interaction between humans and animals obsolete. Second, *intensification* occurs. Not only is the land more intensively cultivated for feed, but the amount of meat produced by each animal body is intensified. Animals become bigger faster. *Specialization* is the third development. Different species of livestock animals are no longer reared together; instead industrialized farms specialize in one type of animal. Additionally, many slaughter facilities specialize in one type of animal. Fourth, *science and technology* assume a leading role. Scientific and technological developments are used to make animals more conducive to living in conditions of confinement; such interventions include the addition of antibiotics to feed to prevent illness and promote growth and the use of artificial insemination to impregnate with the most desirable genes. Finally, industrialized agriculture is characterized by a dramatically *increased scale*. In the case of animal agriculture this is evident in the dramatically increased number of livestock animals on farms, which in later stages of industrialization come to resemble factories more than what farms used to look like.

The first part of this chapter traces the transition to industrial animal agriculture and the adoption of the five characteristics identified above. The subsequent sections tease apart the specific factors that contributed to these developments, focusing specifically on the past several decades. Although individual factors are identified and discussed, it is important to keep in mind that these factors are interdependent and therefore not mutually exclusive. These factors include governmental policies, scientific and technological developments, economic drivers, ideological influences, and environmental considerations. We begin mapping the transition toward industrial animal agriculture by examining the general technological changes that made it possible.

TECHNOLOGICAL SEEDS OF CHANGE

Historians of science and technology highlight the roles that specific innovations play in contributing to periods of social change. In *An Encyclopedia of the History of Technology*, seven technological eras are delineated.[3] These eras are helpful in orienting ourselves with regard to what was transpiring in the background as we foreground the developments in animal agriculture.

The first era dates back to the australopithecenes, approximately ten million years ago, and is marked by the development of the ability among these hunter-gatherers to create tools and to start and control fires. The second era coincides with the shift from hunting to settled agriculture, around 10,000 BCE. During this period, use of the pottery wheel became widespread and metals were used to make tools and ornaments. The invention of the plough in particular shaped agriculture during this era. The third era is known as the first machine age. The valuable developments of this era include clocks, beginning with portable sundials in the eleventh century, and the printing press around 1440. The fourth era marks the beginning of mass production. After 1500, the press made it possible to create the first mass-produced item: coins. In the late eighteenth century spinning and weaving machines enabled further mass production. During this era the factory system of production began to replace craft-based manufacture. The early nineteenth century ushered in the fifth era, wherein steam technology became ascendant. This technology was applied to transportation (particularly railway locomotives and steamboats). The manufacture of steel also began during this period. A new technology introduced in the sixth era, the internal combustion engine, threatened the steam engine. In 1884, the first high-speed petrol engine was developed, and around the turn of the century Henry Ford made his first car using the combustion engine. The seventh era, although unlikely to be the last, is marked by the expansion of power generation. During this era power was distributed to homes and factories. This power set the stage for a later development that would assume a prominent place in the late twentieth century and early twenty-first century—the computer.[4]

These technological developments are intimately connected to the mass production of animals for consumption. The mass-production format perfected in factories and the development of the steam and combustion engines for transportation were particularly formative for industrial animal agriculture and changed the way meat could be produced, transported, and consumed. These developments, particularly in the area of transportation, made it possible to separate the production of animals from the slaughter and processing of animals from the consumption of animals. This distancing has had significant consequences, detailed throughout this book.

Although humans had been consuming animal flesh, in various forms, throughout the seven eras delineated above, organized attempts to intensify the production of animals really did not emerge until the fourth era. The most dramatic changes in the way livestock animals were raised began in the late nineteenth century and coalesced in the twentieth century; however, the seeds of change existed long before that. Clutton-Brock suggests that the shift to modern livestock farming methods were evident in the attempts at "improvement" of farm animals in Europe in the eighteenth century.[5] During that time the demand for meat grew alongside the developing industrial revolution and urbanization. Improving animals was done by artificial selection, with the aim of increasing production to meet the increasing demand. Clutton-Brock argues that at this point in history livestock animals were no longer viewed as individuals; instead, they became commodities to be improved upon. The early attempts at improvement, however, paled in comparison to what was to come.

INDUSTRIALIZATION: 1850 THROUGH WORLD WAR II

As slaughterhouses industrialized and were capable of killing increasing numbers of animals, more animals had to be produced to feed these enormous slaughterhouses and increasing consumer demand. This was not, however, a unidirectional relationship, and it would be incorrect to conclude that the industrial production of animals did not also influence the slaughter of animals and consumption practices. Indeed, these interconnections are addressed throughout this and subsequent chapters.

The industrialization of animal production became evident in western Europe during the mid-nineteenth century. At this time the prices of farm products fell, while the costs of production (e.g., labor, taxes) increased. In an attempt to bridge the widening gap between declining prices and increasing costs, farmers applied technologies emerging from the industrial revolution to farming. This marked the beginning of two processes that would become increasingly apparent over time: the process of unhooking food production from the natural limits imposed by the environment, animal bodies, and the human labor supply; and the process of food overproduction.[6] It took some time for the technological applications to produce enough animals to meet the demand at the time in Europe. In the meantime, western Europe turned to North America to supplement their supply of meat, which was also made possible by technology.

The development of refrigerated ship technology made the export of meat from North America possible, and by 1880 ships were transporting meat from the United States to Britain daily. With increased consumption at home and abroad, more animals had to be produced in North America, and livestock populations were increased dramatically. As a result, by 1900 the rangelands in the United States had been overgrazed. Feedlots were turned to as a solution.[7] Feedlots took livestock out of pasture and concentrated them in a confined space, as illustrated in figure 2, where they were fed grains to fatten them up more quickly in preparation for slaughter.[8] This was one element of the industrialization process, which resulted in dramatic increases in production. The United States was producing $12 million worth of red meat in 1850; by 1920 production was up to $4.2 billion.[9]

The trajectory of the industrialization of poultry production was somewhat different from the production of cattle and pigs. The farming of chickens remained decentralized and dispersed into the twentieth century. For instance, in 1910, the vast majority of farmers (88 percent) had flocks of chickens, and the average flock size was a relatively small eighty

Figure 2. Feedlot in Rayville, Texas, ca. 1928. (Source: The History Center, Diboll Texas. Used with permission.)

chickens.[10] Until the 1920s, raising chickens was not very profitable, and it was mainly under-taken by women on farms as a side business.[11] Production became more concentrated by chance beginning in 1923. In that year, Cecile Steele, who kept a flock of chickens on her family farm in Delaware, mistakenly received five hundred chicks from a hatchery upon ordering fifty. Instead of sending them back she raised them and received sixty-two cents per pound upon slaughter eighteen weeks later. She reinvested some of this profit and acquired one thousand chicks. Within three years her capacity had increased to ten thousand chick-ens. Inspired by the profitability of her enterprise, approximately five hundred other chicken producers emerged in this same county in Delaware by the end of the 1920s. By 1934, the region was raising seven million chickens a year, making it the chicken capital of the world. By 1942, there were ten slaughterhouses operating in the area, killing approximately thirty-eight million chickens per year. The chickens were then packed in ice and sent to markets.[12] The problem was that these newfound efficiencies soon resulted in overproduction.

Between World War I and World War II there was a significant period of overproduction of meat in general: after World War I, the consumption of meat did not rebound as had been expected because many people did not resume their prewar meat consumption habits. Due to the difference between production and consumption, the livestock market crashed the decade before the Great Depression. Bankruptcy consumed one out of every seven livestock farmers and ranchers.[13] Much of the blame for the crash was bestowed upon farmers, and they were pressured to become more businesslike to prevent this from happening again. The business model that bankers, insurance companies, academic institutions, and the government urged them to adopt was one of increased industrialization. This would only exacerbate the overproduction problem.

In the 1930s, in the midst of the Great Depression, the U.S. government introduced the Agricultural Adjustment Acts (1933, 1938) to redress the overproduction in agriculture and raise farm incomes. The mechanisms for doing so included providing price supports for agricultural products and controlling production by having farmers commit to not exceeding acreage limits. In their thorough analysis of these policies, Winders and Nibert conclude that, while these acts were successful in improving farm incomes, they stimulated further production instead of controlling it.[14] This unintended consequence occurred because the policies limited the amount of acreage that could be used for production, but not the amount of the product that they could produce. Farmers therefore intensified production on the amount of land they were able to use so that they could capitalize on the artificially inflated price of what they produced.[15] The strategy was to produce more animals on less land. These policies therefore further accelerated industrialization in the industry.[16] The Depression-era policies of the federal government also facilitated the overproduction of some crops, particularly corn, soybean, and wheat, which was then compounded by the application of new scientific and technological developments that dramatically increased productivity per acre. Further, the government price supports gave farmers more money to reinvest in increasing productivity, notably through mechanization, chemical fertilizers, and hybrid seeds. Between 1925 and 1929, farms were producing on average 26 bushels of grain per acre. This increased to 39 bushels by 1952, and by 1960–64, the average acre was producing 62.5 bushels.[17] This translates into an astounding 240 percent increase in production over approximately a forty-year period.

HYPERINDUSTRIALIZATION: POST–WORLD WAR II

After World War II, a short-term use was found for some of these surpluses: they were exported to Europe for the reconstruction effort. European agriculture was industrializing at this time, so the market for U.S. surpluses did not last long. In anticipation of this market drying up, some farm organizations began working on solutions to the next impending surplus crisis. Restricting supplies was, of course, one potential course of action; however, instead of restricting supply, the farm organizations advocated increasing demand.[18] How would they do that? Well, the president of the American Farm Bureau Federation suggested the following in his testimony before Congress in 1949: "We are interested in trying to develop policies and programs which will avoid burdensome surpluses in feed grains by encouraging the translation of increased feed production into greater livestock production."[19] By feeding more grains (particularly soybeans and corn) to livestock animals and increasing the amount of livestock animals being raised and killed for meat, the grain farmers could increase the demand for their product. It seemed like a win/win solution for grain and livestock farmers, who were already concentrating the production of animals in feedlots and massive barns.

The government responded by implementing policies to promote meat production. One policy involved providing price support for pork. The government also became involved in research to increase livestock production through increased industrialization; the U.S. Department of Agriculture (USDA) and land-grant universities spearheaded the research in this area.[20] Their research pointed to a way to further increase meat production: the widespread use of concentrated animal feeding operations (CAFOs).[21] The benefits seemed endless: more animals could be produced while reducing the sum of money expended on land, feed, and labor.[22]

CAFOs essentially make it possible to meet the barest biological needs of animals while reducing the amount of resources needed and simultaneously increasing production. The basic principle is referred to as *economies of scale*: by increasing their size and level of production companies are able to reduce their costs per unit because fixed costs are spread out across an increased number of units. Convinced of the potential gains, producers began bringing animals, particularly chickens and pigs, en masse into large barns, where they would live their lives without access to the outdoors. (The living conditions of the animals will be detailed shortly.)

The transition to CAFOs had begun in the poultry industry prior to World War II with Cecile Steele's accidental purchase of five hundred instead of fifty chicks for her farm. This large number of chickens required keeping them in a different, more concentrated and confined, way inside large barns. After World War II, policies, research, and governmental extension services at land-grant colleges promoted further industrialization and provided instruction in doing so. The chicken industry in particular was told that they needed to industrialize in order to be competitive with beef and pork. They were advised they would need to make their product "meatier" and cheaper.[23]

In order to make the transition to the CAFO system successful, a variety of actors would need to be involved (although the term *stakeholder* might be more appropriate here than *actor*): the pharmaceutical industry, animal breeders, and the meat processing industry. The pharmaceutical industry was needed because having so many animals confined in close quarters made the spread of bacteria, viruses, and diseases much more likely. Therefore, the administration of antibiotics and vaccines became routine with the adoption of CAFO production. It is

Figure 3. View of CAFO from road. (Source: ©2014 Rochelle Stevenson)

estimated that by the late 1990s, 10.5 million pounds of antibiotics were added to poultry feed alone. Despite these additives, animals still became ill, requiring veterinarians to administer additional medications, only if more cost effective than letting them die.[24] Breeders came to play an important role in producing hardy lines of animals who could grow bigger and faster than their predecessors. These pharmaceutical and breeder interventions were successful in increasing production, which made it necessary to have a slaughtering/meat processing industry capable of keeping up with production.[25]

It was soon evident that CAFO-based production would come at a cost for farmers. They had to give up a great deal of their autonomy, build new barns, and spend more money to purchase preferred stock, drugs, and feed. For some it required entering into contracts with companies in order to pay for their increasingly expensive operations. Under these contracts, the companies retained ownership of the animals and dictated how farmers would raise and feed them. This further eroded the autonomy of farmers and commonly required them to invest in expensive changes to meet their contractual obligations.[26] Due to increased corporate consolidation in the industry, today farmers have few choices in deciding which company to contract with.[27]

There was speculation that these abrupt and significant changes in meat production could be accompanied by negative consequences. In the 1940s, the USDA hired social anthropologist Walter Goldschmidt, now a professor at the University of California, Los Angeles, to study the effects of the growing industrialization in agriculture. He explicated the benefits of traditional farming, which included contributing to good schools, low community crime rates, healthy populations, and an overall high quality of life, and juxtaposed them with what was transpiring in communities where agriculture was industrializing. He concluded that industrial agriculture was having the exact opposite effect of traditional farming. The results of his study, however, were not popularized because the industry found out about it and voiced their displeasure to the USDA, which stifled the findings.[28]

Against the backdrop of stifled critiques of industrialization in the industry, by the 1960s, university extension services were recommending "integration" to further improve the chicken industry. This process would entail integrating everything from the production of chicks through to the slaughtering and meat processing.[29] It is commonly referred to as *vertical integration*; in short, "one corporation comes to own or control virtually every step of production."[30] Integration was promoted as a way to increase efficiency, provide more control over the provision of supplies, reduce costs, and improve quality.[31] The process swept through the industry, and today, all of the chickens commercially slaughtered are produced by integrated firms.[32]

Some of the desired changes in the industry did occur with increasing integration, along with some perhaps unintended consequences. There is ample evidence that the intended reduction in production costs and increases in productivity were actualized. Between 1947 and 1999, the production costs for chickens dropped by almost 90 percent.[33] While production costs decreased, production increased. The weight of chickens at slaughter was up to 5.06 pounds in 2001 from 3 pounds in the mid-1940s.[34] Chickens also reach their slaughter weight faster: today they reach slaughter weight at approximately forty-seven days, half as long as it took in 1940. Not only was the industry producing more chickens faster, it was producing more of them in total. In the twenty-year span between 1982 and 2002, there was a fivefold increase in the number of chickens produced by the top four chicken companies.[35]

The increasingly integrated companies were growing their productivity and reducing costs through investment in technology. These companies were able to outcompete, and often purchase, smaller farming operations that did not have the resources to invest in technology, which further increased the concentration of ownership in the industry. The growing concentration throughout is evidenced by the fact that by 1983 the number of livestock companies was half of what the industry had been in 1945, while simultaneously the total number of animals produced had doubled.[36] Economists refer to this process whereby large companies grow by acquiring competitors as *horizontal integration*.[37] Using the chicken industry once again as an illustrative example, the high degree of horizontal integration is evidenced by the fact that 69.28 percent of production of chicken for meat is controlled by four companies—Tyson Foods, Pilgrim's, Sanderson Farms, and Perdue Farms.[38]

The changes made in the chicken industry were observed by those in the hog industry, where CAFOs became increasingly common after 1960. Pigs became similarly confined inside large buildings or barns, although the process of industrialization and integration differed somewhat from that in the chicken industry. The hog industry had remained concentrated close to the supply of corn in the Midwest, where pig production had historically been located. Then in the 1990s, firms outside of the Midwest, such as Smithfield Farms, entered the industry. Smithfield, headquartered in North Carolina, facilitated vertical integration in the pig industry when it began contracting piglets out for fattening before slaughter with farmers looking for work upon the decline of the tobacco industry.[39] The integration process in the industry progressed quickly: the proportion of pigs being produced through integrated operations increased from 11 percent in 1993 to 59 percent only six years later.[40] Further, the number of hog operations decreased concurrent with an increase in the number of hogs produced—evidence of increasing industrialization. In Nebraska, for instance, between 1965 and 1997, more than thirty thousand facilities stopped producing pigs, while the number of pigs produced statewide increased from 2.8 million to 3.5 million.[41] By 2001, industrialized hog farms (those with more than five thousand pigs) owned 75 percent of the hogs in the United States, up dramatically from 27 percent only seven years earlier.[42] The most recent estimate is that approximately 86 percent of the pork supply in the United States comes from pigs raised in CAFOs with at least two thousand pigs and 61 percent from pigs raised in CAFOs with more than five thousand animals.[43]

The production of cattle has also changed significantly. Grazing cattle on grass used to be widespread. In the late nineteenth and early twentieth centuries, increased demand for beef and expansion of railway transport made the use of feedlots to "finish" or fatten cattle up for slaughter with a diet of grain attractive. This type of production also resulted in soft, marbled meat that consumers wanted. After World War II, corn became the preferred feed. Geography, however, created somewhat of an economic obstacle. The states where the cattle industry was concentrated (e.g., Texas and Nebraska) were not located in the corn belt, and shipping grains to these states was expensive. Instead of rethinking the use of grain to feed cattle, irrigation systems were installed in the cattle states in the 1950s to tap into the aquifers to make it possible to grow grains.[44] Grains consequently became more plentiful and cheaper.

The feedlot system, including feeding cattle a diet of corn and antibiotics, increased productivity. By the turn of the twenty-first century, the feedlot system was bringing cattle to full weight and slaughter by the time they were a year and a half old—half as long as it had taken prior to World War II.[45] Today, most cows spend their first six months grazing on an open range. Then they are shipped to feedlots.[46] Approximately 60 percent of the beef cattle in the United States are contained in 262 feedlots with more than sixteen thousand cows each.[47]

Industrialized CAFOS, or factory farms as they are colloquially referred to, are now ascendant. Ninety-seven percent of the livestock market in the United States heralds from this

Figure 4. Cattle on Florida feedlot. (Source: ©2013 Amy Fitzgerald)

method of production.[48] Farmers have had no choice but to follow the advice given by Earl Butz, secretary of agriculture under President Nixon, to "get big or get out."[49] This development has been framed as progress: "Grazing and pasture are becoming archaic necessities which have been largely replaced by hormones, grains, and concrete."[50] While the preceding material has detailed *how* this transition took place, we now turn our attention to understanding the more proximate factors that have shaped this process. Although there are a multitude of factors, I focus on what I consider to be the most significant ones here: governmental policies, science and technology, economic drivers, ideology, and environmental limits.

PROXIMATE FACTORS SHAPING INDUSTRIALIZED ANIMAL AGRICULTURE

Governmental Policies

According to a working paper by Tufts University's Global Development and Environment Institute, the 1996 farm bill (the Federal Agriculture Improvement and Reform Act) was the most significant legislation vis-à-vis farming since the Depression-era policies.[51] The bill eliminated the supply control programs that had been put in place in the 1930s (discussed earlier). Within four years of the removal of these controls, the production of grains, notably corn and soybeans, had risen dramatically. At the same time, however, the market values of these commodities and the income of the farms producing them stagnated or declined. In response, the government implemented support payments to farmers, which failed to solve the problem; instead, "under the new system, market prices fell below the cost of production, and government spending on farm subsidies went through the roof. In 2000 alone, government aid made up 100% of net farm income in eight states."[52] By 2005, the prices of soybeans and corn had declined 21 percent and 32 percent respectively, while production of each increased by 42 percent and 28 percent.[53]

The authors of the Tufts report ask an important question: Who benefits from these drops in grain prices? Clearly, it is not the grain producers themselves. The authors identify the livestock industry, agribusiness traders, purchasers, and companies such as Monsanto that provide chemical inputs as benefitting from the depressed grain prices. The USDA has acknowledged that the use of these cheap grains as feed has contributed to the increasing integration and industrialization of hog production in the United States and Canada. The grain in the United States is so cheap that many pigs bred and raised in Canada are transported to the United States for fattening in CAFOs because the lower feed prices make it advantageous, even taking the cost of transportation into account.[54]

The authors of the Tufts report argue that subsidization of grains amounts to an implicit subsidy for the animal agriculture industry. They calculate how much chicken production companies gain from purchasing feed as currently valued versus what the cost would be if feed prices reflected the true costs of production to arrive at an estimate of the implicit subsidy. They find that in the decade before the introduction of the 1996 farm bill, corn and soybean prices were 10 percent below actual costs, and the implicit subsidy to chicken production companies was $377 million per year on average, an impressive per year savings. The authors find this savings increased even further to $1.25 billion per year on average after the introduction of the 1996 farm bill. This represents a 230 percent increase in the implicit subsidy. After the introduction of the farm bill, feed prices were on average 21 percent below the actual cost of production.[55]

The authors of the report restrict their empirical analysis to the chicken industry; however, they provide a preliminary analysis of the pig and cattle industries and assert that they are also benefitting from artificially low feed costs. Industrial hog producers feed their pigs a blend of 80 percent corn and 17 percent soybean, and 54–65 percent of their costs are feed. The authors therefore estimate that the implicit subsidy to the industry post–1996 farm bill would be $566.3 million on average per year. The authors conclude that these subsidies create an unfair cost advantage for industrialized animal production over smaller, diversified farms raising animals in free-range environments. The same is no doubt true of the cattle industry, but the authors do not offer an estimate of the implicit subsidy there. The authors further conclude that governmental farming policies are pushing the hog and cattle industries in the direction of the now fully integrated and industrialized chicken industry.[56]

In the long run, these recent agricultural policies have harmed instead of assisted farmers. They have facilitated the growing consolidation and industrialization in the livestock industry and have failed to help grain farmers, whose incomes have been dropping. At the same time, taxpayers are shouldering the increasing costs of subsidies, which have increased dramatically since the passage of the 1996 farm bill. More specifically, corn subsidies increased from approximately $2.1 billion to $4.2 billion in 2008, soybean subsidies increased from $152 million to over $2 billion, and direct livestock subsidies went from $84 million in 1996 to nearly $335 million by 2008.[57]

Science and Technology

Although governmental policies have continued to facilitate the industrialization and concentration in the livestock industry, it would not have been logistically possible without the contributions of science and technology, which have modified the animals' bodies and living environments. One intervention the industry has been working on of late is to increase the uniformity of livestock animals produced by breeder companies to make handling and processing as well as slaughter easier and faster.[58] For instance, the industry has reportedly expressed an interest in breeding chickens without feathers and livestock animals without legs.[59] An even more significant intervention has been proposed in the form of cloning livestock animals, which at least one animal agriculture company, Smithfield, has invested in thus far.[60] The animal breeding sector has been characterized by horizontal integration, as larger animal breeding companies have acquired smaller ones, leading to corporate consolidation. In recognition of their value to the industry, many animal breeding companies have been enveloped by large meat production and processing companies as part of the process of vertical integration.[61]

Although these most recent applications of science and technology to standardize animal bodies may read like science fiction, humans have been attempting to modify livestock animals' bodies for quite some time. In the eighteenth century, scientific interventions with livestock animals were largely in the form of breeding or "improvement" programs. Then in the nineteenth century, those involved with animal husbandry began to record statistical information about the breeds. The twentieth century brought more detailed and intrusive interventions in the lives of livestock animals, and it was at this point that the animal's body itself came to be viewed as a factory.[62] In his book *Animals as Biotechnology*, Richard Twine does an excellent job delineating what has been and continues to be done to animal bodies in the name of increased production and reduced costs. He uses Michel Foucault's conceptualization of *biopower* to explore how power is exerted upon animal bodies as if they were machines and the implications thereof.[63]

In his book, Twine also usefully distinguishes between four types of molecular biotechnology techniques at play in the industry. The first is marker-assisted selection, where markers within an animal's genome are identified and used to guide breeding. The chicken genome was sequenced in 2004, cow in 2009, and pig in 2012.[64] The second is genomic selection, which was developed in 2001 and is currently used in commercial production. This technique utilizes thousands of markers in order to select for rather complex traits. The third technique, which has not yet been used in commercialized animal agriculture, is genetic modification. This technique uses the insertion of novel genetic material into an animal or modifies its existing genes. The final technique is cloning, which has also not yet been used in commercialized animal agriculture. This technique is undertaken by replicating the nuclear DNA of an organism. Twine speculates that the latter two techniques may be utilized in commercial animal agriculture shortly due to a lack of public involvement in decisions about food production and because, as he puts it, regulations are "enabling tools for capitalization."[65] He also points out that neither the United States nor the United Kingdom have made any policy changes in response to the rapidly developing biotechnologies.

The reasons for utilizing these technologies in animal production are varied; the most common include to increase growth rates, to improve disease resistance, and increasingly of late, to mitigate the environmental impacts of the industry.[66] Illustrative of this final goal, a genetically modified pig, named Enviropig, was developed at the University of Guelph in Ontario, Canada, in conjunction with Ontario Pork, an industry trade group. The pig was developed to digest more phosphorus than standard pigs, therefore releasing less phosphorus in manure, which reduces the possibility of phosphorus contamination of the surrounding environment. It also had the added advantage of reducing feed costs.[67] The project seemed like a win/win for the industry and even environmentalists; however, it came to an end in mid-2012 when Ontario Pork reportedly pulled their funding of the program because of "consumer antipathy toward the altered animals."[68] In other words, they were afraid consumers would not buy the pork and it would not be profitable. The project garnered public attention, not because of the scientific and technological developments being experimented with, but because the pigs involved in the project were killed because they could not be released to other farms or sanctuaries because they had been genetically modified.

Despite the purported gains that have been made for the industry due to science and technology, some scientists remain skeptical. They suggest that these interventions could result in behavioral, physiological, and immunological problems in the treated animals and could ultimately exacerbate the already growing problem of reduced genetic diversity among livestock animals.[69] Additionally, the application of technologies to increase production seems somewhat irrational given that production of livestock animals currently exceeds the demand. Further, the application of some scientific and technological developments have worsened animal welfare. For instance, while the productivity of chickens has been increased through genetic manipulation to increase body weight, doing so has contributed to painful physiological problems.[70]

Economic Drivers

Commentators on animal agriculture most frequently describe the situation we find ourselves in today as the result of technological developments in the industry. Gunderson, however, rightly observes that "industrial technologies are the most *visible* transformation of the livestock revolution; however, few confront *why* industrial technologies have superseded old ways of farming and ranching and will, most likely, continue to do so." He goes on to state, "Without understanding the sociohistorical development and drive of capitalism and its unique need to produce for exchange values, industrial technologies in food animal production will continue to be considered an autonomous 'problem' to be dealt with rather than particular instruments of labor to increase profit margins."[71] Gunderson's words are a useful reminder that the scientific and technological developments were implemented primarily because there were economic reasons to do so. Therefore, as Gunderson argues, it is not enough to simply describe the (technological) developments in animal agriculture; we also need to explain these developments.

Although in this chapter a number of factors that have contributed to the industrialization of animal agriculture are delineated, it is worth underlining the fact that the motivation at the root of these changes has been economic: these developments were brought to bear not to further the goal of feeding the masses, but instead were driven by the goal of profit maximization. The best way to demonstrate that this is the case is to review the historical context. The amount of time in our history that has been spent producing food for profit is relatively brief and recent. Before animal domestication humans largely scavenged for meat. Once animals were domesticated they served several functions, including performing labor around the farm, being used as a food source, and sometimes being traded for other subsistence goods if there were surpluses. These functions contributed to the overriding goal of subsistence, not profit.

The surplus of farm animals, as with the surplus of human labor, was later imbued with an exchange value and not just a use value—a process that is said to have emerged in sixteenth-century England. Individuals were no longer producing food directly for their own subsistence; they were producing for money so that they could afford food (among other things).[72] Thus, there was suddenly a reason to produce more meat than one's family could consume, which people did through a variety of means, including producing more animals, producing larger animals, and producing animals who took a shorter time to reach maturity. These techniques were being developed via selective breeding in eighteenth-century Europe.[73] As discussed, the technologies brought to bear on animal bodies became ever more invasive in the twentieth century.

Increasing production is not the only way to increase profits; reducing costs is another means to achieve that goal. This method of profit maximization was accomplished by reducing the costs of raising and caring for livestock animals, such as through modifications to feed and living environment. Increased production and reduced costs merged in the twentieth century in the form of hyperindustrialized animal production. Although the poor conditions in which animals are produced and reproduced that gain public attention are often framed by the media, the industry, and the government as the result of the pathology of individual workers or mismanagement, it is arguably an institutionalized consequence of the two methods of profit maximization: "brutality is *structural*."[74]

Beyond just increasing production and reducing costs to maximize profits, capitalism is constantly seeking to expand and accumulate more.[75] Ever-increasing growth is the goal, which can be actualized through means such as technology and creating new markets; this growth, however, comes with notable environmental consequences. Schnaiberg coined the term *Treadmill of Production* (ToP) to describe this process, whereby the speed of production and use of environmental resources constantly accelerates to the point that it produces its own demise.[76] Novek applies Schnaiberg's ToP concept to industrial animal agriculture and argues that localities court industrial animal agriculture facilities, even when they are aware of the externalities they produce, because of the much sought after growth potential for their local economies—growth that could go elsewhere because the ToP, like corporations, crosses national boundaries.[77] The short-term benefits obscure the longer-term costs.

The impetus to industrialize to increase productivity and decrease costs did not only come from within animal agriculture; other industries played a role. The banking industry contributed significantly to the industrialization of animal agriculture because in the past, and still today, they look more favorably on the industrialized model of animal agriculture and are more likely to grant loans in that direction. As farming became increasingly capital-intensive and the integrator companies became ascendant, farmers have increasingly needed to take out loans to be competitive. In addition to domestic banking policies, broader global trends of financial and labor market deregulation, along with the increasing power of transnational corporations, have also contributed to the industrialization of animal agriculture.[78] Additionally, retailers (such as grocery store and fast food chains) have exercised pressure on producers to reduce costs. And as retailers controlled increasingly more access to consumers the pressure became even more effective.[79]

In combination, these economic factors have contributed to the thoroughgoing industrialization of and consolidation in industrial animal agriculture. Although the contribution of scientific and technological factors to industrializing the industry is more visible, the less visible economic factors were at the base. Even less visible are the ideological factors that have played a role.

Ideology

Although one might not think of ideology and agriculture as being related, Hardeman and Jochemsen demonstrate that agriculture has more ideological elements than we might expect.[80] They set out to examine whether the industrialization of agriculture (in the Netherlands in particular) has ideological traits, and if so, which traits are apparent. In their analysis they search for eight classical and contemporary characteristics of ideology. The first is the presence of narrow-mindedness and an absolutized end. Through their analysis, they find that an absolutized end does indeed exist in industrialized agriculture: efficiency and, by extension, profit. The second ideological characteristic they search for is the reformulation of ideas, norms, and values to be consistent with the ideology. They find that although the Dutch government might not have aimed systematically to change the ideas, norms, and values surrounding agriculture, they did transfer their objectives through informational and educational initiatives with farmers. We might find a parallel to this in the extension efforts of land-grant universities in the United States described earlier. The third characteristic is using the selected end as the only criterion for the choice of method employed. They argue that this is apparent in agriculture where technologies are adopted regardless of whether or not they will lead to efficiencies. We witness this in the form of specialization and intensification. This is indicative of ideological influence because the concern is no longer whether or not technology should be implemented, but how it should most efficiently be done.

They also search for more contemporary characteristics of ideology, such as a well-thought-out rational structure, or rationalization—their fourth characteristic. For evidence of the existence of this characteristic in agriculture they point to the planning undertaken by the government and claims that increasing efficiency in agriculture is rational by nature. The fifth characteristic is a radical depth and breadth (radicalization). They assert this characteristic is evident in governmental programs related to agriculture (e.g., grants, taxes, promotion funds) and the technological changes they fund. These are radical interventions that have literally changed animal bodies and the environment they live in. The sixth characteristic of ideology they search for is the most efficient use of resources, or instrumentalization. A clear illustration of instrumentalization in agriculture can be found in how livestock animals are treated in order to maximize efficiencies and production. What is not used as meat will even be instrumentalized for other purposes, such as to make leather or glue.

The belief that no alternatives to the selected end can be found, and a resulting propensity to attempt to resolve problems using the exact methods that caused the problems, is the seventh characteristic of ideology they identify and search for. As evidence, they suggest that when troubles within the industry do arise, the same means—industrialization—are used in an attempt to resolve them. For instance, when chickens begin cannibalizing each other due to their close living quarters, the solution is to debeak them (see figure 5), not to rethink the industrialized production. The final characteristic they look for is the use of ideological or religious terms in reference to agricultural industrialization. They point to terms used to describe industrialized agriculture in the literature, such as "miracles of modernization" and

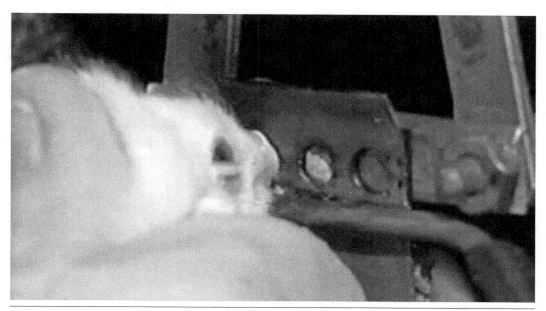

Figure 5. Chicken debeaking. (Source: People for the Ethical Treatment of Animals. Used with permission.)

"spectacular," and explain that this pseudo-religious terminology also supports their contention that the modernization of agriculture possesses ideological traits.

Based on these findings, Hardeman and Jochemsen conclude that the industrialization of agriculture is imbued with significant ideological elements.[81] They suggest that this process likely began with the Enlightenment, when a more instrumental view of nature began to take shape:

> since Enlightenment thinkers considered religion a superfluous source of meaning of life, they proposed new "worldly" goals with their values as a replacement. This harbored the risk of people attributing quasi-religious traits to those ends. If this happens, the end becomes absolutized and an ideology is born. This pattern is . . . found in agriculture to a certain degree.[82]

This is particularly problematic in the instance of agriculture because this ideology keeps us from questioning the process—the ideology reaffirms its taken-for-granted-ness. The progression thesis is a particularly pernicious ideological element vis-à-vis animal agriculture; according to the thesis, animal agriculture *had* to develop to the form it is today and this development is indeed progress. It is assumed that subsistence agriculture is somehow uncivilized and something to be liberated from.[83] Further, the ideology has powerful proponents, such as large agricultural companies.

In order to pull the curtain back on animal agriculture ideologies, Barbara Noske introduced the concept of the animal-industrial complex.[84] Richard Twine recently wrote a paper seeking to revive the concept as a way to further the dialogue between critical animal studies and political-economic approaches and usefully provides the following definition therein: the animal-industrial complex is "a partly opaque and multiple set of networks and relationships between the corporate . . . sector, governments, and public and private science. With economic, cultural, social and affective dimensions it encompasses an extensive range of practices, technologies, images, identities and markets."[85] Looking at animal agriculture through this lens is helpful because it foregrounds the capitalist impetus behind modern industrial animal agriculture (and other forms of animal oppression), it draws attention to the multitude of actors in this complex (e.g., animal agriculture companies, retailers, pharmaceutical companies, government agencies, educational institutions), it acknowledges that the oppression generated by the animal-industrial complex is tied up with other complexes in our culture, the impacts of which expand beyond nonhuman animals,[86] and it challenges the normalization of industrial animal agriculture. This normalization is supported by ideology. As Noske states, the "animal industry . . . , by claiming to serve the basic human interests of health and well-being, largely succeeds in getting itself regarded as (sadly) inevitable."[87] This notion that the current mode of industrial animal agriculture is an inevitable consequence of progress limits critical examination of these developments. In this context, ideologies that facilitate production in the name of efficiency and profit while instrumentalizing the animals and rationalizing their treatment continue largely unabated.

Environmental Considerations

Environmental pressures have also contributed to, and likewise been exacerbated by, the growth and concentration of livestock production. Overgrazing made confinement methods using grains as feed attractive, and now seemingly necessary: global livestock populations currently exceed the environmental capacities of grazing land. As geographer Tony Weis explains, "Because global livestock populations now far exceed stocking capacities on rangelands and pasturing on small, integrated farms, the continuing meatification of diets cannot happen without the continued growth in factory farming and the cycling of industrial grains and oilseeds through livestock."[88] In other words, the high and growing demand for meat and current (in)availability of suitable grazing land makes the CAFO system the most efficient, and arguably necessary, way of producing animals for human consumption. In this way, demand does clearly contribute to the current mode of production. Consumptive demands are detailed in a later chapter.

At the same time that environmental pressures have contributed to the industrialization of animal agriculture, industrialized animal agriculture threatens the environment, which is detailed later in the book. Suffice it to say for now that industrialized animal agriculture contributes to deforestation, air pollution, water pollution, and global climate change, to name a few environmental impacts. Further, the ability of science and technology to mitigate the threats against the environment is uncertain. As illustrated in the previous discussion of science and technology, although the industry is working on some technological fixes to reduce their environmental impact, such as in the case of EnviroPig, such technologies will only be brought to bear if they are deemed profitable, and in that case they were not because consumers were not interested in consuming EnviroPigs. To make matters worse, the pressures on the environment are increasing as developing countries are attempting to replicate the western diet laden with animal products, which is discussed in the next section. These developments are, in turn, furthering the ascendancy of the industrialized CAFO system.

THE GLOBALIZATION OF INDUSTRIALIZED
CAPITALIST ANIMAL AGRICULTURE

As discussed earlier, the capitalist ethos is expansion. Corporations are under pressure to continually expand, and if they fail to expand they risk losing the faith of their shareholders. For instance, as I write this book the shares of technology giant Apple Inc. dropped 10 percent because their first quarter earnings had not increased as expected. This cost Apple tens of billions of dollars in market value. Interestingly, Apple was still experiencing dramatic profits ($13 billion in the quarter), but their growth had slowed somewhat.[89] This illustrates that profits are not sufficient: ever-increasing profits are the goal.

One might reasonably ask, How can a company continue to increase their profits once they have seemingly exhausted the consumer market domestically? One of the answers is to expand into untapped international markets. By the turn of the twenty-first century, the meat industry had saturated the market in the United States, so they began expanding their horizons internationally in two main ways. First of all, they began exporting meats produced domestically to international markets. For instance, meat products from the United States are exported to the Philippines, where the products are sold at prices cheaper than what the Filipino industry can (thus far) produce them. Under the General Agreement on Tariffs and Trade (GATT) and the World Trade Organization (WTO), the Philippines cannot impose tariffs on these products.[90] Even if the Philippines wanted to keep animal agriculture products produced through industrialized means out of the country, they would face an uphill battle. For instance, some countries have used international trade agreements to argue that the animal welfare standards required in other countries, particularly members of the European Union, amount to trade protectionism.[91]

The second way in which the U.S. meat industry is entering new markets is by setting up livestock production and processing facilities in developing countries. For instance, since the 1980s, Smithfield, Tyson, and Cargill have set up operations in China, and they sell their products there.[92] Burmeister hypothesizes that in the future it could become financially beneficial for industrial animal agriculture to follow the trend in other industries and outsource parts of the production abroad and then sell the final product domestically.[93]

In the interests of expanding the new markets further, the meat industry has also been promoting increased meat consumption in developing countries. Whereas Twine refers to the increasing consumption of meat by the sizable populations in developing countries as a "sustainability time bomb," for the companies involved it is a potential jackpot.[94] Since 1950, there has been more than a fivefold increase in meat production globally.[95] In the past two decades alone, the consumption of meat and dairy products in developing countries has tripled, and egg consumption has quadrupled.[96] These trends are being produced by simultaneous increases in human population and in per capita meat consumption.

The global livestock population is already so large it is almost difficult to comprehend. At any point in time there are 17 billion chickens (they outnumber people by a ratio of 3.5 to one), 1.3 billion cows (which translates to one for every five people), over one billion sheep (one for every six people), one billion pigs (one for every seven people), and 800 million goats (one for every eight people).[97] These are just the most populous animals; there are others that are consumed such as turkeys, rabbits, and horses. The number of land animals killed globally every year is approximately fifty-six billion.[98] Reflective of the increased meat consumption in developing countries, a growing share of livestock animals globally are located there. See table 1, which

Table 1. Production of Meat Products by Region, 1980 and 2007 (in million tonnes)

Region, Country Group, or Country	1980	2007
DEVELOPED COUNTRIES	88.6	110.2*
Former centrally planned economies	24.6	19.0
Other developed countries	64.0	91.3
DEVELOPING COUNTRIES	48.1	175.5
East and Southeast Asia	19.4	106.2
China	13.6	88.7
Rest of East and Southeast Asia	5.6	17.5
Latin America and the Caribbean	15.7	40.3
Brazil	5.3	20.1
Rest of Latin America and the Caribbean	10.4	20.2
South Asia	3.7	9.4
India	2.6	6.3
Rest of South Asia	1.1	3.0
Near East and North Africa	3.4	9.7
Sub-Saharan Africa	5.5	9.3
WORLD	136.7	285.7

Source: FAO, "State of Food and Agriculture."
Note: Totals for developing countries and the world include a few countries not included in the regional aggregates.
*Total sum may be off due to rounding.

provides an overview of the amount of meat produced in developed and developing countries. It is estimated that by 2020, 60 percent of the livestock animals in the world will be located in developing countries. This trend makes what transpires in animal agriculture in developing countries particularly important. Industrialization in animal agriculture is already taking hold in these countries; globally thus far, approximately half of the cattle live in CAFOs and more than half of the pigs and chickens live in CAFOs.[99]

These increases in meat consumption and industrialized animal production are not simply attributable to corporate efforts. For years, international development agencies have been championing industrialized production of crops and livestock animals in developing countries. Only recently has the World Bank adopted a more cautious approach to funding industrial animal agriculture projects in developing countries due to concerns about the displacement of people, environmental degradation, and food (in)security.[100]

Developed countries have also been contributing to increased production in developing countries. As developed countries have degraded their own resources and land has become increasingly scarce, these nations have been looking to less developed countries, particularly those with weaker environmental and animal welfare laws, for agricultural land, reminiscent of when Europeans colonized North America.[101] Sociologist David Nibert does an admirable job in his book *Animal Rights/Human Rights: Entanglements of Oppression and Liberation*, and in a journal article coauthored with Bill Winders, of demonstrating how shifting industrial animal agriculture to developing countries is a form of neocolonialism.[102] The consequences of this neocolonialism include the dislocation of subsistence farmers and indigenous populations, decimation of forests (specifically in Central and South America), further consolidation of landownership in the hands of the few, and brutal military crackdowns on resistance. China is currently exploring agricultural options in the Congo, Cambodia, Laos, Indonesia, Zimbabwe, and Mozambique.[103] Mexico, India, countries of the former Soviet Union, and some Asian countries are already home to intensive production operations.[104] Mexico, for instance, is attractive because there are no governmental restrictions on the size, concentration, or location of CAFOs.[105]

Industrial animal agriculture has been promoted as a method for bringing prosperity and food security to developing countries. What has happened instead is that, as these countries have increasingly adopted intensive industrial animal agriculture methods, food security has deteriorated because food production has become concentrated in a few geographic areas, leaving some locations without, and oftentimes different breeds of livestock animals are used in industrialized operations than were used in subsistence farming, which can threaten the native breeds. For instance, in the Philippines, the breed of chicken used in their massive and growing poultry industry is not native to the region and is responsible for decimating the native chicken breed through the spread of diseases foreign to them, thus compromising a local food source.[106]

As a result of these and other problems, cautions about industrial animal agriculture are increasingly being directed at developing countries, which are being urged not to follow the development path charted in the United States and many other developed countries. For instance, Deborah Fitzgerald warns that

> development experts ignore at their peril the complex history of industrial agriculture, for the patterns of change set in the American case are now replaying all over the world. The benefits of industrial agriculture are invariably countered by the costs, whether those are measured in financial, environmental, or human terms.[107]

I would add the cost of increased animal suffering to this list. Matheny and Leahy argue that, instead of continuing to weaken extant food systems by trying to compete with capital-intensive livestock production in developed countries, developing countries should instead capitalize on a built-in strength they possess: inexpensive labor.[108] It would cost less than in developed countries to produce more labor-intensive, high-value, free-range livestock. They could therefore capitalize on the fact that animal welfare and environmental sustainability can add value to meat products in some markets.

Shifts in production have not only been occurring from developed to developing countries, but have also taken place from developed countries to other developed countries. Recently some companies have begun moving livestock operations from western Europe to the United States, due to more lax environmental regulations there (regulations on the disposal of manure, for example, are more lenient in the United States).[109] So, if some companies are transferring their livestock production to the United States, motivated at least partly by the lax regulations, this begs these questions: What regulations on industrialized animal agriculture are actually in place in the United States? And how do they compare to other jurisdictions?

THE REGULATION OF INDUSTRIAL ANIMAL
AGRICULTURE PRODUCTION IN THE UNITED STATES

In general, the federal government has failed to adequately regulate CAFOS, choosing instead to leave much of the regulation up to the states. Some federal laws not specifically designed for CAFOs, however, have been applied to them in an attempt to mitigate the harms that have become increasingly apparent. For instance, livestock waste was added as an area of concern to the federal Clean Water Act, although the application of the act to CAFOs has been muddied by legal challenges and changes.[110] The federal Clean Air Act has been even less effective in regulating CAFOs, particularly because the Environmental Protection Agency (EPA) excluded odors from the act, thus limiting the ability of citizens to sue over CAFO odors (which have demonstrated negative effects on property values and quality of life). The EPA does have the ability to censure CAFOs for excess particulate matter in the air, but to date the agency has not demonstrated much of a willingness to pursue this route.[111]

Most of the regulations that have been implemented to specifically govern CAFOs are at the state or municipal level. A few states have developed more stringent CAFO regulations, oftentimes after a significant public health or environmental incident. For instance, Iowa now requires that CAFOs be sited a minimum distance from homes and wells, regulates the construction of manure lagoons, requires that CAFOs submit manure management plans annually, and requires that those handling manure for commercial purposes be licensed by the state and meet certification and education requirements. Wisconsin requires CAFOs to annually analyze the manure they apply to the land to make sure the pollutants are within permissible limits, and they are required to test the land surrounding the CAFO every four years. And in 2002, Minnesota placed a moratorium on the construction of CAFOs.[112] Burmeister argues state- and municipal-level regulations have been reactionary and emerged as responses to the globalization of neoliberal policies, particularly deregulation, in an attempt to mitigate the localized impacts of industrial animal agriculture.[113] Be that as it may, these regulations have been better than what other jurisdictions have been able to implement.

Kentucky was thwarted in its attempts to impose regulations on the poultry industry. Motivated by state subsidies (in the form of tax breaks, assistance with infrastructure development, and worker training), weak unionization in the state, limited land zoning regulations, and access to grain for feed, the poultry industry was drawn to Kentucky. By the turn of the twenty-first century, the state was producing over two hundred million birds, which made it the thirteenth largest producer in the country. After witnessing the environmental degradation in neighboring North Carolina as a result of CAFOs, the governor of Kentucky authorized the temporary regulation of CAFOs by the state. The regulations were to include requiring CAFOs to be sited a significant distance away from houses, schools, and recreation areas, requiring that they secure permits for water and waste discharge, and that integrator companies would be jointly liable with individual farmers they had contracted with for the creation of environmental hazards. The Kentucky Farm Bureau filed a lawsuit over the constitutionality of these temporary regulations. After significant legal wrangling the power of the temporary regulations was thwarted.[114]

Other states were not as proactive in at least attempting to regulate CAFOs and instead courted their business. For instance, Nebraska rolled out the red carpet for the pig industry and was successful in attracting them. Chase County, Nebraska, had a population of 5,600 pigs in 1997; within four years they had more than 80,000 pigs. Citizens voiced concerns about the expanding CAFOs. Then in 1998 and 1999 when the Nebraska legislature took up the issue of CAFOs they acted in favour of the industry because they were concerned that regulating it, particularly trying to mitigate the odour, would have economic repercussions for the state.[115]

Taking protection of the industry one step further, some states, such as Missouri and Michigan, passed "right to farm" laws that protect CAFO owners from lawsuits as long as their actions fall within "generally accepted agricultural management practices" (GAAMP).[116] In Michigan, the Sierra Club sued the Department of Environmental Quality over the state's (lack of) regulation of CAFOs. The court of appeals determined that part of the state's CAFO regulations were not consistent with the Clean Water Act. Nevertheless, any gains were offset by the fact that the state has "right-to-farm" laws. Therefore, "CAFO manure spreading, even in excessive amounts, is considered a GAAMP, despite the pungent odor and the known environmental impacts. Thus, suits brought by private citizens for the most common CAFO problems are dismissed with nothing done to rectify any pollution caused by the CAFO."[117]

There has even been evidence of corruption vis-à-vis regulation of the industry. In 1995, the *News and Observer* newspaper in Raleigh, North Carolina, ran its "Boss Hog" series exposing how the growing pig industry in the state was being aided instead of regulated by the government, and putting citizens at risk as a consequence.

> In a seven-month investigation, The *N&O* found that state agencies aid the expansion of pork production but are slow to act on a growing range of problems resulting from that increase. The industry has won laws and policies promoting its rapid growth in North Carolina. It also has profited from a network of formal and informal alliances with powerful people in government.[118]

The series detailed the environmental degradation taking place, corporate consolidation in the industry, and the relationship between the pig industry and government officials.

The reporters focused on Wendell Murphy, who became the largest hog producer in the nation during the ten years he served in the state General Assembly. His farm had been the first to industrialize pig production in the 1960s. He developed an integrated operation where he provided his contract farmers with the feed and the pigs for them to grow. Then he took the animals back when they were full grown while the farmers kept the manure. His farm is also credited with designing a building to confine pigs all day based on the model of chicken CAFOs.[119] The reporters accused him of helping "pass laws worth millions of dollars to his company and his industry. The Duplin County executive voted for, and sometimes co-sponsored, bills giving hog and poultry producers tax breaks, protection from local zoning and exemptions from tougher environmental regulations."[120] The consequences of the laws passed include exempting CAFOs from zoning regulations, gutting penalties for illegally discharging manure, and exempting building materials for CAFOs from sales tax. Ironically, Murphy's actions did not actually violate the state ethics law as long as he could claim that the money he stood to make from the votes did not affect his judgment.[121] The newspaper subsequently won a Pulitzer Prize for public service for publishing the "Boss Hog" series of stories.

Some states not only allow the industry to conduct business as usual, they further protect it by prohibiting public criticism of CAFO-produced food. Eleven states have such laws on the books.[122] It may be tempting to shrug these laws off as toothless or unenforceable, but they are not. It was under this type of law in Texas that TV personality Oprah Winfrey was sued for stating she would no longer consume beef after the "mad cow" outbreak. Further, U.S. citizens are not just vulnerable to such disparagement laws in the United States. Robert F. Kennedy, a vocal opponent of industrial animal agriculture, claims that for years he has been followed by an industry employee who records his speeches, and in fact, after speaking out against industrial animal agriculture in Poland, which has relatively weak freedom of expression laws, he was charged there criminally.[123]

More recently, there has been a move in some states to criminalize taking photographs or film footage inside CAFOs. In response to undercover footage documenting animal cruelty in CAFOs and slaughterhouses, "rather than vow to enforce humane standards of treatment, lobbyists for America's factory farms have instead introduced legislation in states across the country, including California and Vermont, to essentially punish whistle-blowers."[124] Such "ag-gag" laws, as they have come to be referred to, have already been passed in Iowa, Missouri, and Utah. The *New York Times* published an editorial denouncing ag-gag laws and warned about the consequences, stating,

> Their only purpose is to keep consumers in the dark, to make sure we know as little as possible about the grim details of factory farming. These bills are pushed by intensive lobbying from agribusiness corporations and animal production groups. The ag-gag laws guarantee one thing for certain: increased distrust of American farmers and our food supply in general. They are exactly the wrong solution to a problem entirely of big agriculture's own making.[125]

The industrial push for ag-gag laws in particular begs the question: What are they trying to hide? We turn now to an examination of the conditions inside CAFOs to address exactly that question.

CONDITIONS IN CAFOS

Because the production of animals for meat has become so removed from its consumption, very few consumers have actually seen firsthand how the animals live, much less how they die. It is a difficult thing to do justice to via the written word. The first time I saw inside a CAFO was years ago when I was a teenager, and I likely would not have even known what the term "CAFO" meant at the time. A friend of mine worked at a chicken CAFO, and he told me they were going to be rounding the birds up in the evening to be sent to slaughter; they did so under the cover of darkness because the birds are more compliant when they cannot see well. He explained the process to me rather blankly, but something changed in his demeanor when he began to tell me about what was done with the "runts" who were not physically "good enough" to go to slaughter. The employees would "play baseball" with them: killing them by using them as "balls" thrown to a "batter." This bothered my friend; it was easier to be detached from the chickens that were being sent to slaughter in the distance, but this was clearly cruelty and it was more proximate. We decided that I would go to the farm the evening they were being rounded up, and I would take the runts.

The evening came, and I was nervous. I did not know what to bring to put the chickens in to transport them in my car. Truth be told, I had never really interacted with a chicken before and did not know what to expect, but in my mind I envisioned angry chickens that would harm me if they got the chance—and I wouldn't blame them, given how they had been living.

From where I was parked I could see inside the barn as human figures worked together to round up the birds in the dim light. My friend emerged, seemingly from out of nowhere, with one chicken in his arms. He informed me this was the only runt tonight. What he brought me looked more like an alien than a chicken. She had very few feathers and was very small and thin and scratched up. We placed the bird in the laundry basket I had brought and put a towel across the top so she could not escape. Little did I know she would not have the energy to escape. I pulled away quickly. When I got home with the chicken I quietly took her in the laundry basket into the garage. I did not know what to expect when I took the towel off the laundry basket to let her out. She didn't fly out of the basket, as I had worried; instead I had to lift her out. To my surprise, she let me and continued to let me pet her.

I quickly began to think about what to name her. I settled on the name Marian, as I was at that time reading Margaret Atwood's book *The Edible Woman*, wherein the protagonist, Marian, begins to critically reflect upon the consumption of animals.[126] Marian followed me around the yard and actually came when I called her better than my dog did. I wondered if she remembered where she had been and how far she had progressed physically. Over time she put on weight and more feathers grew in—she began to look less like an alien and more like a chicken. When I had to do a presentation in my high school English class I decided to focus it on the theme of vegetarianism in Atwood's work and brought Marian to school with me for it. I also brought my cat and guinea pig because I wanted the students to see how Marian was just as friendly as my other "pet" animals. I remember a very surprised teacher and fellow students who actually paid attention that day.

I share this story because it was my first glimpse into how animals are raised for slaughter and ultimately our plates. Unfortunately, most people will never have an opportunity to interact with an animal from a CAFO or see what these facilities are like on the inside. People really ought to have information about where their food comes from, but the industry is making glimpses such as the one I got increasingly less likely. I doubt that today, twenty years later, I would be able to drive right up to a CAFO and take a chicken, even though the chicken was going to be brutally killed and not used for food. In fact, today I would risk being charged and labelled as a terrorist.

The vast majority of the population has to rely on secondary sources to find out what goes on inside CAFOs, if they seek to find out at all. There have been a number of documentaries that have given viewers some idea, such as *Food Inc.*, which although being fairly sanitized pulls back the curtain on CAFOs with some footage inside chicken-growing operations. *Meet Your Meat* and *Earthlings* provide undercover footage, but viewers often find these more difficult to get through. I encourage you to watch what you can because seeing it is more vivid and authentic, I think, than any description I or anyone else could write, although I think award-winning author Jonathan Safran Foer does a commendable job of giving a vivid account.[127]

Foer gained access to a few CAFOs and does an excellent job describing what goes on within those walls. In his book *Eating Animals*, Foer describes what he saw when he crossed the threshold into the CAFO world, and he escorts the reader through the experience:

> Try to picture it. (It's unlikely you'll ever get to see the inside of a poultry farm in person, but there are plenty of images on the Internet if your imagination needs help.) Find a piece of printer paper and imagine a full-grown bird shaped something like a football with legs standing on it. Imagine 33,000 of these rectangles in a grid. . . . Now enclose the grid with windowless walls and put a ceiling on top. Run in automated (drug-laced) feed, water, heating, and ventilation systems. This is a farm.[128]

In what follows I detail the living conditions inside such "farms." The scene being described depends heavily on the species contained in the CAFO, but one thing is the same across species in industrial animal agriculture: they are treated as machines,[129] which serves to implicitly deny their sentience.[130] These are the realities.

Their environment is determined by the drive to maximize production above all else.

Chickens and pigs are kept in large barn-like, windowless buildings. From the outside, one would have a difficult time envisioning what is going on inside these buildings. They are rather nondescript, although they generally have large, tell-tale ventilation fans on the sides (see figure 6). Chickens used for meat (referred to as "broilers") are crammed on the floor of these buildings (twenty thousand in one building at some facilities) and have to fight their way to get feed and water. To make matters worse, they have been bred to grow very large very rapidly, and oftentimes they suffer bone and leg injuries due to their rapid growth and can even be afflicted by "acute death syndrome" where they suddenly jump into the air and flip over dead. Broiler chickens live in their own manure, and that of other chickens, on the floor of the barn. Due to the cramped environment it is standard practice to debeak them (pictured earlier) without anesthetic when they are young because they will peck and even cannibalize each other in such close confinement, which would result in reduced production.[131]

In 98 percent of the facilities that produce eggs the chickens are packed into cages on top of each other (four to five chickens occupying the space of a folded newspaper). Some of these facilities house two hundred thousand chickens in one building. Many chickens used for egg

Figure 6. CAFO, with ventilation fans pictured. (Source: ©2014 Rochelle Stevenson)

production suffer from "caged layer fatigue," where the constant depletion of calcium for egg production leads to deteriorated bones and muscles and commonly an inability to reach food and water, resulting in death. Unlike "broiler chickens," chickens used for egg production do not have to sit in their own manure; it runs through the bottom of the wire cage they are in. However, they instead have to stand on the wire cage, which causes pain and deformity in their feet. In hatcheries, where chickens are being bred to repopulate the laying chickens that are slaughtered once they become fatigued, the male chicks are killed because they are not useful to the industry. They are killed by suffocation, gas asphyxiation, or being ground up by blender-like machines.[132]

Pigs are also commonly confined indoors (see figure 7), although the form their confinement takes depends largely on their life stage. Approximately 83 percent of female breeding pigs give birth in total confinement. They begin in what is referred to as the prefarrowing area, where they are kept for breeding (the majority through artificial insemination) and gestation. For the duration of the four-month gestation period pregnant pigs are kept in "gestation crates," which are so small that they cannot turn around. This confinement prevents the pig from grooming herself or engaging in other natural behaviors such as nesting.[133] These crates are so restrictive that at the time of this writing, nine states are phasing in bans on their use.

Shortly before or immediately after they give birth, the pigs are moved to the farrowing area where they are kept with their piglets in tight confinement for nursing. The mother is given only enough room to eat, drink, and nurse. Again, the crate she is kept in is so narrow that she cannot turn around. Shortly after birth, and without anesthetic, parts of the piglets' teeth and tails are cut off, and males are castrated. As with debeaking chickens, these procedures are performed because the piglets will harm each other otherwise due to the extremely close quarters.[134] In some facilities, at least, runt piglets are reportedly killed by swinging them by their hind legs and smashing their heads on the floor.[135] After weaning, surviving piglets are separated from their mother and go into finishing pens, or sometimes to different farms that specialize in fattening pigs, where they are kept until they put on enough weight to be sent to slaughter. The mothers go back to the prefarrowing area and keep reproducing until they are "spent," at which time they also go to slaughter.[136] Like chickens, some pigs never make it to slaughter because they die suddenly from "porcine stress syndrome." Pigs currently bred in the industry for rapid growth are vulnerable to this condition.[137] Pigs raised in CAFOs also have high rates of pneumonia due to the constant inhalation of the ammonia in their waste.[138]

The majority of cows, and not just those raised for meat, are also kept in confinement. Most dairy cows are kept in extreme confinement, tethered in stalls. Like the other animals used in CAFOs, their lot in life is worsened by the fact that they have been bred to produce—they produce high volumes of milk, which contributes to physical problems, particularly painful inflammation of the udders.[139] In order to keep cows producing milk they must be continually impregnated. Some of the resultant calves, particularly the males, are then used in the veal industry. Approximately 750,000 calves per year "are taken from their mothers within a day of birth and turned into sickly, neurotic animals to provide the luxury-grade 'milk-fed' veal preferred by gourmet cooks and fancy restaurants."[140] These calves are kept in extreme confinement, in very narrow stalls and tied at the neck, for eighteen to twenty weeks. They are fed a milk-replacer with less iron than their mother's milk would contain in order to make their flesh white due to subclinical anemia. The anemia induces them to lick their own urine or feces as a source of iron, but the extreme confinement keeps them from being able to reach that potential source of iron.[141]

Figure 7. Industrialized pig production. (Source: ©The Humane Farming Association. Used with permission.)

"Beef" cattle are commonly dehorned and castrated without anesthetic, which the industry claims only causes stress. "Cattle are fully conscious when blades gouge out their horns, slicing through arteries, veins, and nerves. During castration a calf or bull feels the knife slit his scrotum and sever his testicles."[142] The majority of "beef" cattle are fattened up on feedlots on a diet of grains instead of grazing on grasses. This unnatural diet often leads to liver abscesses in cattle, and the cramped quarters on the feedlot lead to stress.[143]

As discussed above, sickness among the animals in CAFOs is endemic. As Jonathan Safran Foer puts it, "Needless to say, jamming deformed, drugged, overstressed [animals] together in a filthy, waste-coated room is not very healthy."[144] Deaths are common, but even with these losses, CAFOs are an economical way for companies to produce animals. Among other cost savings, this confinement reduces the amount of land needed because more animals can be crammed into closer quarters; it reduces the amount of feed needed, because the animals are so confined that they burn fewer calories due to restricted movement; and it results in reduced labor costs because the concentration makes monitoring easier and less labor-intensive, as does the mechanization of feeding and watering.[145]

These are the realities of industrial animal agriculture, which most people will never witness. As discussed in this chapter, there is not one factor that pushed us to this type of production; it is more complicated than that. A number of factors—including governmental policies, technological developments, economic drivers, ideology, and environmental issues—all contributed to this transition, with the influence of each fluctuating over time. Industrialized animal agriculture is not only now ascendant in the United States, it is gaining momentum in developing countries. As a result, not only has the face of production changed, but processing has changed dramatically as well.

The Industrialization of Slaughter and Processing

THE SLAUGHTERING OF ANIMALS IS AN ESSENTIAL COMPONENT OF TURNING PRO-
duced animals into food, or what I refer to in this chapter as *processing* animal bodies into
consumables. As Noëlie Vialles points out, meat can only be obtained through slaughtering animals, as those who die beforehand are deemed unfit for consumption (although, as
discussed later in this chapter, that does not necessarily preclude them from being put into
the food supply).[1] Despite the essential role played by animal slaughter and processing more
generally in bringing meat to the table, as Vialles observes it is an "unpopular subject." (And
writing about animal slaughter is complicated by the knowledge that few people want to read
about it.) Further, changes in the industry, which will be discussed in this chapter, have made
it even easier for consumers not to think about the slaughterhouse at all.

Language also makes it easier for people to avoid thinking about how and where livestock animals are killed. The word *slaughterhouse* was originally used to refer to any building
where animals were slaughtered for human consumption.[2] More recently, the word has been
somewhat replaced by more sanitized terms, such as *meatpacking plant* and the French word
abattoir. The National Cattleman's Association has even suggested avoiding using the verb
slaughtering and instead prefers *processing*, *harvesting*, and *going to market*.[3] In this chapter,
I primarily use the terms *slaughterhouse* and *slaughtering*, as I have chosen to do elsewhere,
because of its clarity in describing what transpires within the confines of those walls: animals
are slaughtered; only then can "meat" be "packed."[4]

Richard York, an environmental sociologist, recommends the development of a "sociology
of the slaughterhouse," which he describes as "a deeper understanding of how exploitation
and oppression (not to mention horrific acts of cruelty) are perpetuated and justified."[5] I agree
with his assessment, and I think that in order to understand the current process of animal
slaughtering and its impacts, and move toward the development of a sociology of the slaughterhouse, it will be necessary to appreciate the historical trajectory that has brought us to this
place—a place where approximately ten billion livestock animals in the United States alone
are slaughtered each year. It is difficult to even fathom the sheer size of that population; to put
it into perspective, it is the equivalent of over three hundred animals being slaughtered every
second.[6] This chapter maps this trajectory beginning with industrialization, which really was,
as they say, a game changer.

The development of the slaughterhouse as an institution has been tied to the production
of livestock animals. As the production of livestock animals became more concentrated and

commercialized, the slaughterhouse emerged as a unique and specialized industry. In this chapter, we first explore the emergence of the slaughterhouse as an institution, followed by an examination of its developmental trajectory. We do so through three major developmental periods. The first period, which grew out of the eighteenth century, is marked by the transition of animal slaughtering from town centres and out of the view of the public. It predates the industrialization of the industry but is necessary for understanding how and why slaughter and processing industrialized. The second major period was ushered in by the industrialization of slaughterhouses, which began in the late nineteenth century. The ultimate example of this period is the Union Stockyard in Chicago. More recently, beginning in the second half of the twentieth century, the third major period took shape as slaughterhouses were relocated from large urban areas, such as Chicago, to small rural communities.[7] Technological, geographic, and workforce changes forever altered the industry.

THE CREATION OF A SEPARATE ANIMAL SLAUGHTERING INSTITUTION

When livestock production was diffused, much more so than it is now, so was animal slaughtering. Families that raised animals for their own consumption commonly slaughtered them in their own backyard. For instance, many colonists in North America kept and slaughtered their own pigs, generally prior to the winter. The meat was then cured so that it would last through the winter.[8] Due to their size, the slaughter of cattle in backyards was less common, and due to the difficulties that accompanied their larger size, home slaughter of cattle was eventually banned in some areas, such as New York as early as 1747.[9]

There was a gradual transition to institutions that specialized in slaughtering animals, not only due to the difficulty of doing so at home, but for other reasons, including social, geographic, and economic. In western Europe and the North American colonies, social reformers agitated to have the "morally dangerous" work more contained.[10] Increasing urbanization also made the raising and slaughtering of animals by individual families less practicable in both regions.[11] Simultaneously, the growing concentration of raising livestock animals made small-scale animal slaughtering less feasible.

As early as 1640, separate seasonal slaughter facilities were established in Springfield (Massachusetts), New York City, Cincinnati, Louisville, and Chicago. In some of these locations, the slaughtering of animals became extremely concentrated. For instance, a stream on Manhattan Island came to be referred to as "Bloody Run" during the prewinter slaughtering season because of the large volume of animals being slaughtered there. In response to the volume, beginning in the 1650s, regulations were imposed on the slaughterhouses, and Manhattan began requiring permits to slaughter animals.[12] The first commercial slaughterhouses emerged around 1660.[13] The commercialization of animal slaughtering gained momentum in the early nineteenth century with the development of internal transportation networks (e.g., roads, canals), which made it easier to get animals to a central location.[14]

In Europe, the transition to specialized facilities for slaughtering animals commenced in the eighteenth century. During this time, social reformers began urging for the relocation of animal slaughtering from private facilities (e.g., backyards, small butchers' sheds) to "public slaughterhouses," where the slaughtering of animals could be concentrated, more easily monitored, and concealed from the general public.[15] Part of their argument hinged on the assertion that "morally dangerous" work should not be left to individual citizens and instead should be actively regulated by the state.[16] New language was needed to refer to this new institution. For instance, the French word *abattoir* was introduced to refer to buildings devoted to killing animals for human consumption.[17]

THE DEVELOPMENTAL TRAJECTORY OF THE SLAUGHTERHOUSE

The Transition of Animal Slaughtering Away from the Public Gaze

The growing concentration of animal slaughter may have been more expedient and allayed concerns about individual citizens slaughtering animals, but it also created a new set of concerns. These specialized facilities slaughtered increasing volumes of animals within urban areas, and citizens began encountering the realities of mass animal slaughter. Although many members of the public had witnessed animal slaughter on a small scale in the past, observing concentrated, mass animal slaughter is unique: As Vialles points out, "whereas the slaughter of a few animals may be a festive occasion, slaughter on a large scale is different. It is disturbing; therefore means must be found of putting it out of mind."[18] Various techniques for putting it out of mind were employed and coincided with a more general cultural desire to eschew the animality from humanity.

Increasing exposure to mass animal slaughter was transpiring against the backdrop of what sociologist Norbert Elias describes as a "civilizing process" that the western world underwent between the sixteenth and nineteenth centuries.[19] As part of this process, some ways of interacting with food became shameful. More specifically, many people tried to distance themselves from the production of meat. Keith Thomas documents how, beginning in the eighteenth century, individuals tried to separate themselves from the slaughter of animals, both rhetorically (e.g., by claiming that s/he could never do the work of butchers) and literally (e.g., by erecting barriers between their property and slaughter facilities).[20] He also details a prejudice against butchers that took shape in the seventeenth century. The butcher was cast as an evil character, instead of as someone who was merely doing the work demanded of him by consumers. This prejudice was no doubt related to the desire among the populace to distance themselves from animality, and animal slaughter by extension.

The case of the Smithfield Market in London is illustrative of the concerns about slaughterhouses in town centers at this time and the agitation to have them removed.[21] The Smithfield Market had been opened in the tenth century and was a site of animal slaughter and sale (the immense size of which is illustrated in figure 8), which had grown over the years. Concerns about the market increased over time, and by the late eighteenth century the market was being monitored by authorities and a licensing process for slaughterhouses was put in place.[22] In the nineteenth century, concerns about the market coalesced. The concerns were primarily around the risks to citizens in the form of morality and disease.

In the mid-nineteenth century, a committee began investigating the market and its impacts. Citizens who were interviewed by the committee voiced concerns about the threat posed by slaughterhouses to morality in the form of alcohol and debauchery, as well as concern that the violence inflicted upon the animals could graduate to violence against people. For instance, Philo cites a shop owner respondent who told the committee that "the chief trades 'encouraged

A BIRD'S EYE VIEW OF SMITHFIELD MARKET,
TAKEN FROM THE BEAR & RAGGED STAFF.

Figure 8. Thomas Rowlandson's *Bird's Eye View of Smithfield Market, London.* (Source: Museum of London/The Art Archive at Art Resource, New York. Used with permission.)

by the existence of Smithfield' were 'gin shops and public houses.'" Another man described a fear that the violence visited upon the animals in the market would "educate the men in the practice of violence and cruelty, so that they seem to have no restraint on the use of it."[23] There were also concerns about the public health risks of slaughtering and processing animals in the city center, which were solidified during a cholera outbreak in the 1840s. As a result of these concerns, the live animal portion of the market was closed in 1855.[24] Animal slaughtering in England was moved to large, "public" slaughterhouses outside of the city center.

Violence toward animals was increasingly being expelled from public view (although artists occasionally brought the realities of the business to the public's attention; see figure 9), and the movement of animal slaughter outside of town limits was one instance of this larger trend. Concomitantly, people were adopting more idealized views of nature and more affective attitudes toward (at least some) animals. This gave rise to a tension, which Thomas articulates as follows: "This was the human dilemma: how to reconcile the physical requirements of civilization with the new feelings and values which that same civilization had generated."[25] This dilemma was being exacerbated by increasing urbanization, which on the one hand created a further distance between people and the production of food, and on the other hand, a developing interest in the animals and nature that people were geographically being separated from.[26]

Parallel movements of slaughterhouses out of city centers were occurring in other western European countries.[27] Although these movements did not unfold in identical ways in different

Figure 9. Annibale Carracci's *The Butcher's Shop*. (Source: Kimbell Art Museum, Fort Worth, Texas/Art Resource, New York. Used with permission)

countries, they were united in the objective of removing animal slaughter from the public's gaze. The French word for slaughterhouse, *abattoir*, came to reflect this distancing of the slaughter-house from the town center. Vialles explains that the definition of the word post-1863 morphed into a "place set aside for the slaughter of animals such as bullocks, calves, sheep, etc. that are used for human consumption. *Abattoirs are located outside the surrounding walls of towns.*"[28] In the English speaking world these slaughtering facilities erected on the outskirts of town were referred to as "public slaughterhouses." The irony is that these "public slaughterhouses" removed slaughter from the sight and mind of the public.[29] In addition to being reflected in linguistic changes, the movement of slaughtering away from the public's gaze had the added effect of improving the reputation of butchers, who were distanced from the slaughter of animals. A new labor force for slaughtering animals replaced the butcher, and they would become associated with the "dirty work" of slaughtering animals.[30] Butchers became less associated with slaughter and increasingly associated with skilled cutting and trimming.

In the North American colonies, concerns about slaughterhouses being located in densely populated areas began to emerge in the seventeenth century.[31] By the 1840s, the slaughter-houses in New York City (approximately two hundred at the time) came under the scrutiny of the urban public health movement. The main concerns about the slaughterhouses included blood and refuse being absorbed in the soil, the emission of foul odors, and the visible transportation of animal body parts through the city. The Health Department in the city eventually required the slaughterhouses to move to less urbanized areas near the waterfront.[32]

Changes also took place in the distribution process. Meat sales to consumers had been restricted to public markets through licensed butchers. As the cities grew, shoppers increasingly complained about the distance they had to travel to purchase meat. Despite protest from butchers, cities legalized the sale of meat in private stores in 1843. This "presented great opportunities to transform the means for supplying larger numbers of Americans eager to obtain a piece of steak" and also marked a separation of slaughter from retail and consumption.[33]

Thus, during the nineteenth century, the United States witnessed the development of increasing gulfs between the processing, distribution, and consumption of meat (out of sight, and increasingly out of mind). Most consumers no longer heard, saw, or smelled the slaughter and processing of animals, and increasingly, fewer came into contact with large dismembered animal body parts in butcher shops. This new invisibility had the effect of neutralizing much of the unease that was percolating about the slaughter and consumption of animals. The geography and architecture of the new nondescript public slaughterhouses also served to assuage a growing "collective cultural guilt" and "had the effect of quelling the potential disruptive potency of this violence to call into question the naturalization of human/animal hierarchy."[34] This emergent separation of production from processing, from distribution, from consumption became institutionalized in the industrialization of the slaughtering process.

The Industrialization of Animal Slaughtering

The separation of the public from the slaughter of animals that began with the exile of slaughtering to the outskirts of towns progressed to a hyperseparated state as animal slaughter became industrialized.[35] Cronon identifies the industrialization of animal slaughter as particularly significant because it marked a break with nature: it severed consumers from the animals, the act of killing, and the natural environment.[36] Vialles also considers this transition significant and observes, "From this point on, slaughtering was required to be industrial, that is to say large scale and anonymous; it must be non-violent (ideally: painless); and it must be invisible (ideally: non-existent). It must be as if it were not."[37] This section of the chapter explores the process of industrialization and factors that contributed to it, including increasing geographic and financial concentration in the industry, technological developments, and growing demand for commercial meat.

During the second half of the nineteenth century, commercial animal slaughter became concentrated in and around a few urbanizing areas, specifically Cincinnati, Chicago, New York, St. Louis, and Kansas City.[38] Cincinnati earned the title of pig slaughtering capital. In the 1830s, one hundred thousand pigs were slaughtered in Cincinnati; that number increased fourfold by the time the Civil War commenced.[39] During the Civil War, however, the level of pork processing in Cincinnati began to be eclipsed by the slaughtering undertaken in Chicago.[40] New York held the title of cattle processing capital until the 1880s when it was also eclipsed by the growing slaughterhouse powerhouse of Chicago (see figure 10).[41] Chicago became dominant because it had two key advantages: it was more integrated into the growing railroad network and was located closer to the supply of livestock animals than the cities in the Northeast.[42] These advantages were exploited by the Union Stock Yard and Transit Company.

Figure 10. The Union Stockyards complex in the Packingtown district of Chicago. (Date: 01/01/1919). (Source: NGS Image Collection/The Art Archive at Art Resource, New York. Used with permission.)

Although the first slaughterhouse was erected in Chicago in 1827, it was not until 1865 that production capacity increased dramatically, facilitated by the incorporation of the Union Stock Yard and Transit Company.[43] The Union Stockyards took up over a square mile of land and employed at least five thousand people.[44] It became quite a spectacle, and souvenirs emerged, such as the postcard in figure 11. Many of the workers lived in a slum in the back of the yards; the area was characterized by extreme poverty, crowded conditions, delinquency, and environmental pollution. The deplorable working and living conditions cultivated prounion sentiments. The community grew until the 1940s and ultimately became home to nearly sixty thousand people, approximately half of whom were immigrants.[45]

The Implementation of Refrigeration Technology

By 1875, the introduction of cooling technology in slaughterhouses made slaughtering year round possible because the meat could be safely stored.[46] The application of this technology to meat transportation further revolutionized the industry. George Hammond, who ran a slaughterhouse in Detroit, was reportedly the first to develop a system for transporting frozen meat. The profits he earned from this endeavor are said to have inspired Gustavus Swift to get into the beef shipping business.[47]

In 1878, Swift and Andrew Chase developed a more reliable refrigerated railcar that came to be referred to as the Swift-Chase car. This technological development made it possible to slaughter and process animals a greater distance away from the market and then ship the refrigerated consumable parts (which came to be referred to as "dressed beef") to market.

Figure 11. The Union Stockyard, meat packing district, Chicago, Illinois. Postcard, USA, ca. 1900. (Source: CCI/The Art Archive at Art Resource, New York. Used with permission.)

Shipping only the parts of the animals that were marketable reduced shipping costs, which was significant because a large proportion of carcasses is not consumed. For example, approximately 60 percent of cattle bodies are unsellable.[48]

Chicago was embedded in the developing railway network (see figure 12) and therefore well poised to capitalize on the new refrigerated railcar technology. The refrigeration technology made Chicago the capital of animal slaughter and processing and "had the effect of nationalizing beef, as people no longer had to buy locally produced or recently slaughtered animals."[49] It could be sent from a central location as needed. Consumers were at first skeptical about the safety of consuming an animal that had been killed a week prior, but the lower price won them over and demand grew.[50] By World War I, refrigerated routes reached twenty-five thousand communities in the United States.[51] The growing ascendancy of animal slaughter and processing in Chicago, however, was not without its critics. Some states resisted the growing dominance of Chicago by implementing laws prohibiting the importation of dressed beef. These laws were eventually struck down by the Supreme Court on the grounds that they unfairly restricted interstate commerce.[52]

The effects of refrigeration technology were profound—so much so that the implementation of the technology has been described as "annihilating space."[53] Not only did it facilitate the rise of Chicago as a meatpacking powerhouse, it made the processing of larger amounts of meat possible because meat no longer had to be sold or canned immediately after slaughter. Increased supply meant reduced costs. The technology therefore ushered in "a new 'era of cheap beef,'" which altered not only the type of meat people were consuming but also the amount.[54] This technology also further solidified the separation between processing and consumption, as cuts of meat could be prepackaged and sent greater distances to consumers, and livestock production

Figure 12. The extensive network of railroad tracks throughout the stockyards allowed workers to unload livestock directly into pens. (Source: Chicago History Museum, photographic print ICHi-13877, ca. 1959, photograph by John McCarthy. Used with permission.)

and processing, because it made it possible to locate animal slaughter a further distance away from livestock production.[55] (Figure 13 captures the loading of cattle on to railcars, an increasingly common scene at this time.)

The trajectory the evolution of chicken slaughter took was somewhat different than the just described path that pig and cattle slaughter took. In the nineteenth century, farmers brought chickens, other poultry, and eggs with them to urban markets. When the weather was cold they brought slaughtered and dressed poultry to market; during warm weather they would bring the poultry to market live and then kill each as purchased. The introduction of refrigeration technology made it possible to mass slaughter and process chickens away from the markets and also took the seasonal variation out of production.[56] Yet this technology did not facilitate the high degree of slaughter concentration for chickens as it did for pigs and cattle, primarily because the production of chickens was not yet concentrated:

the widespread availability of chickens on farms and corresponding dispersion of slaughtering and dressing among urban centers precluded centralization of the poultry industry paralleling beef and pork. Instead, as demand grew so too did highly regional chicken markets, as adjacent farming regions learned to supply the needs of growing urban centers.[57]

Figure 13. Cattle to and from Rayville [Louisiana] traveled by rail. Rayville siding, ca. 1928. Note the Texas and New Orleans stock cars and Southern Pacific railroad stock cars. (Source: The History Center, Diboll, Texas. Used with permission.)

It was not until the mid-twentieth century that the slaughter of chickens became concentrated geographically (discussed in the section on "Changing Workforce Composition and Organization").

Introduction of the (Dis)assembly Line

Another technological development—the mechanized slaughterhouse disassembly line—also dramatically changed the slaughtering of livestock. This technology appears to have been first introduced in the hog slaughter sector of the industry in the United States as early as the 1850s. The scene at an early disassembly line in Armour Company's facility in Chicago was captured in a wood carving, pictured in figure 14. The disassembly line continually moves animal bodies between work stations so that workers do not have to take the time to physically move them themselves. It increases production dramatically, as illustrated by the fact that in 1884, five "splitters" were required to process 800 cattle in a day. Yet only ten years later, the mechanized disassembly line made it possible for only four splitters to process 1,200 cattle a day.[58] This technological development "ushered in the industrial mantra that 'time is money,' as packers became more and more focused on finding the most efficient and cost-effective ways to dismantle animals into their constituent and marketable parts."[59] The lesson that fewer workers and increased production translated into increased profits was shared with other industries.

The disassembly line not only revolutionized the animal slaughtering industry, it also inspired changes elsewhere. Henry Ford used the disassembly line in slaughterhouses as the model for the assembly line he developed in 1913 for building cars.[60] Rifkin and Shukin point out that the formative role that the meat processing industry played in industrial innovations

Figure 14. One of the earliest production lines: Armour Company's pig slaughterhouse, Chicago. Pigs walked up the ramp to the top of the building, then were processed and emerged as finished carcasses. Eviscerating carcasses. Wood engraving, Paris 1892. (Source: Universal History Archive/UIG/Bridgeman Images. Used with permission.)

and the fact that it was the origin of Fordism as a unique socioeconomic philosophy have been overlooked, and attention instead has focused on the role of the steel and automobile industries in industrializing production.[61]

In addition to this broader socioeconomic impact, the disassembly line changed the working environment inside slaughterhouses dramatically. The technology made it possible to further divide the labor up into very specific, discrete tasks. This in turn reduced worker control over the skill and time needed to perform their assigned task and increased the speed of production.[62] Subsequent mechanical innovations in the early twentieth century, such as the application of mechanized saws, further sped up meat processing.[63]

The scope of the disassembly line eventually expanded and incorporated work that used to be done outside of the slaughterhouse. In particular, slaughterhouses integrated trim work that had been done externally by butchers. Prior to the twentieth century, large cuts of meat were sent from slaughterhouses to butchers, who would then do the more precise work of transforming sides of meat into the specific cuts that local consumers wanted. The meat processing companies began to undertake this trim work inside the slaughterhouses, and initially employed butchers to do so. Due to their skills, butchers earned a higher wage than general slaughterhouse employees. Looking to cut costs wherever possible, the slaughterhouses eventually replaced butchers with unskilled workers and used the disassembly line to do the work instead; by 1910, butchers had been nearly entirely eliminated from slaughterhouses.[64]

The dramatic division of labor ushered in by the disassembly line, which has in a sense now been perfected, also had the effect of diffusing the blame for animal deaths that had historically been placed on butchers. With different people now restraining the animals, stunning them, slitting their throats, and disassembling the carcass, identification of who actually takes the animal's life became less clear.[65] The virtual removal of butchers from this equation resulted in the transfer of the previous prejudice directed at butchers for the work they did onto slaughterhouse workers en masse.[66]

The Environment inside Industrializing Slaughterhouses

In addition to its productive qualities, the disassembly line became a spectacle, and a profitable one at that. The companies operating slaughterhouses with the early disassembly lines not only used them to produce meat, they also used them to attract visitors. The companies gave public tours of their facilities and proudly displayed the speed and efficiency that industrialization had made possible. The slaughterhouse "became a symbol of this country's industrial capitalist might," and the companies were happy to put it on display.[67] One company even published a souvenir guidebook.[68]

Shukin argues that these tours were significant because they were a precursor to the cinema in that they utilized physical displacement as entertainment. They were also significant because they provided a window into the production of meat—a window that has been firmly closed in more recent years. Many members of the general public attended these tours and witnessed what was transpiring inside the slaughterhouses in person. Others learned about the realities of animal slaughter through the written word. For instance, author and poet Rudyard Kipling toured the stockyards and described the physical environment to readers as follows: "There was no place for hand or foot that was not coated with the thickeness of dried blood . . . and the stench of it in the nostrils bred fear."[69] The experience apparently made quite an impression on Kipling. According to Cronon, "indifference seemed to Kipling the most frightening thing he saw at the stockyards, and made him worry about the effect of so mechanical a killing house on the human soul."[70]

Another author wrote an entire novel based on the conditions in the slaughterhouses in Chicago. Upton Sinclair aptly titled his book *The Jungle*. Sinclair reportedly had first sought to publish the manuscript with MacMillan Company; they agreed to publish it as long as he sanitized it by removing the "blood and guts" from the pages. He refused and sought different publishers but received several rejections. He was so committed to getting his story out that he eventually published it himself. Shortly thereafter, Doubleday, Page and Company agreed to publish it as long as they could verify the contents. An editor visited the Union Stockyard slaughterhouses in Chicago where Sinclair had based the book and verified the story. The book was finally published and eventually became a best seller, capturing the attention of the general public and the government.[71]

In the book, Sinclair chronicled the working conditions in the stockyards, such as dim lighting, poorly ventilated buildings, poor sanitation, and workers standing in water, blood, and feces. Injuries, infections, and disease, such as tuberculosis and pneumonia, were rampant, and medical attention sparse.[72] Sinclair also chronicled horrifying incidents of workers being accidentally dismembered and killed, and human remains ending up in the meat supply.

Sinclair reportedly wrote *The Jungle* because he wanted to expose the horrible working conditions inside slaughterhouses. In doing so, he simultaneously exposed the realities of how meat was being made, which instigated a public outcry and resultant attention by the government. President Theodore Roosevelt even invited Sinclair to the White House to discuss the horrors he had recorded on the pages of his book. Then in 1906, Roosevelt signed two pieces of legislation into law in an attempt to redress the problems with meat production detailed in Sinclair's book. The Pure Food and Drug Act banned interstate trade in adulterated or mislabelled food and drug products, and the Federal Meat Inspection Act expanded inspection from hogs to all livestock (with notable exceptions, including poultry, fish, and game) being exported across state and international lines, and also gave the USDA increased regulatory power.[73] Against this backdrop of exposés, public outcry, and legislative actions, however, the meatpacking industry was becoming increasingly profitable and powerful.

Growing Concentration of Power in the Industry

From the second half of the nineteenth century into the early twentieth century, the economic value of the animal slaughter and processing industry increased significantly. In 1850, the industry in the United States was valued at $12 million. By 1920, it was valued at $4.2 billion, it was the second largest employer nationally, and it was one of the largest contributors to the gross national product.[74] In short, the animal slaughtering industry was becoming an economic powerhouse.

At the same time that the value of the industry was growing, control within it was becoming increasingly concentrated in the hands of a relatively small number of companies, notably Armour, Morris, Cudahy, and Schwarzschild and Sulzberger (later Wilson & Company). These companies came to be referred to collectively as the Beef Trust. By 1916, these companies controlled 82 percent of cattle, 77 percent of calf, 61 percent of hog, and 86.4 percent of lamb and sheep interstate slaughter.[75] This degree of concentration of power in an economic sector can be referred to as an *oligopoly*, which is defined as relatively few companies supplying a large number of customers. Skaggs contends that the industry could also be described as an *oligopsony*, which is characterized by the existence of many sellers (e.g., livestock producers) and a few buyers (e.g., the slaughtering companies). This may have been an accurate description at the turn of the twentieth century, but the number of livestock producers declined markedly in the second half of the twentieth century, and vertical and horizontal integration became the norm. Therefore, while the term *oligopoly* (and some have even argued the term *monopoly*) could apply to the industry today, the term *oligopsony* would now be less appropriate.

The Beef Trust companies not only became known for the enormous amount of power they wielded in the animal slaughtering and processing industry; they also received unwelcomed attention for their questionable business dealings. For instance, in the 1860s, attention was drawn to Philip Armour and Gustavus Swift when they conspired with competitors to fix prices. They were also known for producing tainted meat and repressing attempts at unionization among workers.[76]

These types of business dealings, coupled with the high degree of corporate concentration of ownership in the industry, attracted the attention of the federal government by the end of World War I. At this time, the Federal Trade Commission stated that, in retrospect,

> since the middle 1880s the "beef trust" had systematically conspired to manipulate livestock markets; restrict interstate and international supplies of food; control the prices of dress meats and other foods; defraud both the producers of food and consumers; crush effective competition; secure special privileges from railroads, stockyard companies, and municipalities; and profiteer.[77]

The companies that comprised the Beef Trust were eventually charged with antitrust violations; however, the charges were suspended in 1920 when the companies agreed to a consent decree—a binding agreement between parties engaged in a lawsuit. Under the decree, the slaughterhouse companies had to relinquish their holdings in other business sectors, such as livestock production, stockyards, railroads, and newspapers, among others.[78] In response to this increased scrutiny and pressure to alter their business practices, the American Meat Packers Association (later changed to the American Meat Institute) was created by the industry to lobby on its behalf.[79]

Soon after the consent decree was signed, the companies and their lobby group began to complain that as a result of their divestments in other business sectors they were losing the market to smaller competitors in the slaughterhouse industry. The decree was revised to allow the companies to own refrigerator cars and trucks twelve years after it had been agreed to. The decree was later further revised to enable the companies to manufacture and distribute nonmeat products at wholesale.[80] After these revisions, the concentration of ownership in the industry rebounded. Illustrative of this resurgent concentration, by 1935, one company alone—Swift and Armour—controlled over 61 percent of national meat sales. Admitting that the watered-down governmental efforts had been for nought, in 1957, a congressional report stated that the concentration of power in the animal slaughtering industry had reached the level it had been prior to the issuing of the consent decree.[81] This high degree of consolidated power in the industry has constituted a formidable hurdle for workers in their attempts to improve the working conditions inside slaughterhouses from the nineteenth century to the present day.

Unionization in the Industry

Frustrated by not having their concerns addressed, slaughterhouse workers mounted strike actions in the late nineteenth century. The companies they were up against, however, were able to mitigate the impact of striking employees by bringing workers from other plants, predominantly immigrant and black workers. Violent incidents erupted with the use of these replacement workers; for instance, several of the black strikebreakers were hanged by angry striking employees.[82] Federal troops were even brought in to Chicago to quell violence that erupted when the newly formed Stockyard Butchers' Union engaged in a strike in 1894 against the largest five slaughterhouse companies.[83]

Two years later, the Amalgamated Meat Cutters and Butcher Workmen of North America (AMC) union was formed. Within a few years of its inception, industry-wide agreements were negotiated. There were numerous setbacks, however, in the subsequent years. Negotiations with the meatpacking companies broke down in 1904, and the union's fifty thousand members went on strike. Again, replacement workers were brought in and riots ensued. The companies proved to have more staying power and were able to outlast the workers, who eventually gave up their strike, and some even gave up on their union. A second nationwide strike occurred in 1921. This time fifty-five thousand workers walked off the job. Again replacement workers and police were called in, and within a couple of weeks the union had lost the strike. It was not until the late 1930s that they were able to negotiate lasting agreements with the large slaughterhouse companies, thanks at least partially to new labor relations legislation.[84]

The introduction of the National Labor Relations Act in 1935 meant that the meatpacking industry had to negotiate in good faith with the unions and was prohibited from engaging in a list of unfair labor practices.[85] By the end of World War II, the United Packinghouse Workers of America (UPWA) had negotiated national contracts with the large slaughterhouse companies, and within approximately twenty years, 95 percent of slaughterhouse workers employed outside of the southern states were represented by the UPWA or the AMC.[86] These unions helped to make slaughterhouse work one of the best paying industrial jobs, although the working conditions still remained relatively dangerous. The gains made by the unions at this time, however, were rather short lived. By the end of the 1960s, the unions began losing ground.[87] Several factors contributed to the decline of unions in the industry (detailed shortly), including a geographic shift in animal slaughtering—the third and most recent developmental period in the slaughterhouse industry.

TECHNOLOGICAL, GEOGRAPHIC, AND WORKFORCE CHANGES

It is impossible to pinpoint exactly when this next period in animal slaughtering in the United States began. There was not one specific event that marked the beginning of this new era; instead there were a series of minor changes that in and of themselves would not have seemed all that significant, but in combination they indicated that a wave of change was washing over the industry.

After World War II, new technologies and designs adopted by the industry increased productivity even more. The increases in productivity, however, did not just benefit the corporate giants in the industry; they also benefitted newcomer companies. One new company in particular, Iowa Beef Packers (IBP; changed in 1970 to Iowa Beef Processors),[88] was gaining market shares previously held by the large companies that had reigned supreme for years.[89]

It has been asserted that IBP's emergence in 1961 and subsequent growth forever altered the landscape of the meatpacking industry.[90] According to Rifkin, IBP began shaping the industry as early as the 1960s. In the 1970s, the changes it was driving in the industry became increasingly evident. The older companies began to see their profits decline. IBP and other newer companies, such as Momfort and Excel, began to lead the way in the industry and introduced significant changes, including new forms of cutting and transporting beef, a geographic transition of plants from urban to rural areas, increased use of automation, faster line speeds, the elimination of longstanding union contracts, and recruitment of immigrant workers.[91] These interdependent changes transformed the industry into what it is today and are discussed in turn below.

Changes in Meat Cutting and Transporting

When the industry was newly industrializing, discussed in the previous section, it was determined that it was more efficient to locate slaughterhouses in central urban areas and to transport carcasses instead of live animals to market. One of the changes implemented in the most recent era in slaughtering has been to remove the fat and bone, then vacuum-pack and box the meat up. Beef packaged this way came to be referred to as "boxed beef." This made shipping more efficient and less expensive because the uniform packages of meat maximized space during transportation and only saleable pieces of meat were transported.[92]

The precut meat shaped consumption and further cut the butcher out of the processing and distributing processes because it was trimmed in the slaughterhouse and consumers could now purchase the product at the supermarket instead of at a butcher shop. As sales of boxed beef rose, many butchers were put out of business. The move to precut meat also made some of the older slaughterhouses obsolete because they could not compete with the increasing mechanization of newer slaughterhouses, which were better able to respond to the growing demands for precut meat.[93] In tandem with these developments, the channels from production, to processing, through to distribution to consumers were becoming increasingly narrow.

The Transition of Meat Processing from Urban to Rural Areas

It was not only the way the postslaughter product was shipped that changed; shipping animals to slaughter also changed. The dominant practice up to this time was to ship livestock from rural areas to slaughterhouses that were located in urban centers. This process, however, resulted in some losses: there was "shrinkage" or loss of body mass during shipping to the slaughterhouse, bruising, and crippling.[94] These losses were reduced in the most recent era with the relocation of slaughterhouses to rural areas, closer to the supply of livestock, and then shipping the meat to market. This relocation was also attractive because livestock tended to be raised in areas where land was cheaper and in states with low levels of unionization, notably with right-to-work laws, so opening slaughterhouses in these areas also reduced land and labor costs.[95] Responding to these push and pull factors, the meat processing company Cudahy began the transition out of Chicago in 1954.[96] By 1970, the Union Stockyard slaughterhouses in Chicago were closed.[97] The same number of slaughterhouses that had existed in cities, however, did not reemerge in rural areas. Taking advantage of "economies of scale," fewer yet larger slaughterhouses were opened.[98]

Slaughterhouses were relocated to towns that were generally quite small. For instance, most of the towns where IBP relocated their slaughter operations had populations of less than twenty-five thousand.[99] The new slaughterhouse facilities were large but also nondescript buildings; thus, "facing outward, this industrial slaughterhouse blends seamlessly into the landscape of generic business parks ubiquitous to Everyplace, USA."[100]

Changing Workforce Composition and Organization

These small towns did not have large pools of labor available, and the companies seized upon the opportunity to recruit immigrant workers for less pay.[101] The communities where the new, large slaughterhouses opened consequently underwent significant changes. For instance, Garden City, Kansas, where IBP opened a new slaughterhouse in 1980, experienced a 33 percent increase in population within five years, and most of the new residents were southeast Asian refugees and Latinos.[102] This recruitment strategy saved the companies money on wages and made unionization less likely.

To protect their newfound lowered labor costs, IBP and its competitors virulently guarded against unionization, employing both legal and illegal means to achieve this and other ends. For instance, in the 1970s, the *Washington Post* accused IBP of paying off gangsters and employing illegal pricing policies. The cofounder of the company was convicted in 1974 on charges of conspiring with the mafia to gain entry into New York's market. Then in 1987, the company was fined $2.59 million for falsifying safety records and documentation about workers' injuries.[103]

Despite their attempts to thwart unionization, some facilities successfully unionized. Strike actions followed. For instance, the United Food and Commercial Workers Union went on strike in 1982 at an IBP facility in Dakota City. During this strike the Nebraska National Guard was called in. Strikes followed in other cities as unions attempted to retain what gains they had been able to secure in the industry at mid-century; this time they were up against mounting socioeconomic and political forces.[104] Earlier in the century unions had been able to secure wages in cattle and pig slaughter that had at least produced a living wage—the highest in the manufacturing industry. Due to a confluence of factors, by the 1990s, earlier gains were eroded, the unions had little power left, and wages in the industry had dropped.[105]

Several factors contributed to the decline of union power in the industry. Brueggemann and Brown identify three main contributory factors: *economic restructuring, working-class fractionalization,* and *employer ascendancy.* It is worth examining each factor in some detail here because it provides a more complete picture of how unionization in the industry was undermined and insight into the operation of the industry more generally.

The elements of *economic restructuring* in the industry that most significantly impacted unionization include job deskilling due to increasing automation, an increasingly mobile and difficult workforce to organize, and relocation of the industry to regions with histories of weak unionization, already discussed.[106] It should also be noted that high turnover in the industry has also posed a formidable obstacle to unionizing, and in fact it has been asserted that the companies like the high turnover rate in the industry (as high as 200 percent per year) because it mitigates against union organizing: most people are not there long enough to work toward organizing the workforce.[107]

Unionization has also been complicated by *working-class fractionalization,* which has intensified in the industry as companies shifted to recruiting more women and racial/ethnic minorities.[108] Racial and ethnic minorities are now the majority working inside slaughterhouses. The most recent statistics (for the year 2012) indicate that the meat and fish processing labor force is comprised of 13.5 percent Black or African Americans, 7.9 percent Asian, and 41.6 percent Hispanic or Latino workers.[109] The industry has also recruited an increasing number of immigrant workers, the majority of which are Latino. Over one quarter of employees in the industry are foreign-born, compared to 10 percent in the rest of the manufacturing sector.[110] This statistic is no doubt a conservative estimate because it excludes undocumented immigrants.

There have been occasional crackdowns on illegal immigrants in slaughterhouse communities. In 1999, the Immigration and Naturalization Service (INS) launched Operation Vanguard, targeting illegal immigrants working in slaughterhouses. This was the first time the INS had targeted a specific industry. Company executives were advised which of their workers were being summoned to the INS for "discrepancies" in their documentation. Out of 24,000 Nebraska and Iowa slaughterhouse workers, 4,500 had such discrepancies. Since this large operation, smaller crackdowns have occurred "to appease the re-emergence of community anxieties over too many immigrants."[111] In order to circumvent the problems with immigration, yet retain this valuable pool of laborers, the industry continues its attempts to get slaughterhouse workers included in guest-worker visa programs, which they have thus far been excluded from because their work is not seasonal.[112]

Immigration crackdowns have had the unintended consequence of drawing the attention of the general public and politicians to the working conditions in slaughterhouses. In response to greater awareness of the working conditions, the Republican governor of Nebraska even had a "Meatpacking Workers Bill of Rights" drafted and ordered that it be posted at the entrance of every slaughterhouse in the state.[113] The changing demographics in slaughterhouse communities has also drawn the public's attention to the industry in a way that otherwise would likely not have happened.

The growing diversity in the rural towns where large slaughterhouses have opened has been resisted in some cases, and some towns have experienced moral panics about "outsiders."[114] This panic is illustrated in an excerpt from Gouveia's field notes from ethnographic research in a slaughterhouse community, where she recorded one respondent as stating, "This town is ruined since the Mexicans came. It used to be that you could leave your car unlocked or your children's bicycle on the front yard. These people don't even take care of their lawn."[115] Due to their marginalization and precarious positioning in the labor market, both documented and undocumented immigrant workers are less likely to agitate for improved wages, benefits, and unionization.[116]

The power of unions has also been thwarted by what Brueggemann and Brown refer to as *employer ascendancy*, which is related to the high degree of integration and consolidation in the industry.[117] The size and power of employer companies in the industry has increased dramatically, which is particularly significant given the power the employer companies have had even historically. A brief examination of integration and consolidation in the chicken sector of the industry, the most integrated of all sectors, serves to illustrate the challenges posed by employer ascendancy.

Chicken slaughter did not become concentrated as early as it had for cattle and pigs. Until the mid-twentieth century, chicken production and processing remained decentralized and took place on a rather small scale. This decentralization was facilitated by the fact that, due to their smaller size, chickens are relatively easy to slaughter and dress on one's own property, and their meat can be consumed rather quickly and does not therefore require storage.

It was in the Delmarva peninsula that mass production and processing of chickens first became meaningfully concentrated. It was the concentrated production of chickens in this region that facilitated the development of large, concentrated slaughterhouses. By 1942, thirty-eight million chickens were being processed annually by ten slaughterhouses in the region. The plants were originally similar to the nineteenth-century pig slaughterhouses in their methods. The small size of the chickens and their relative uniformity, however, made it relatively easy to adopt technologies to assist with the processing, such as pulling out the feathers, and made the work less labor intensive. By 1945, the technologically developed facilities could process one hundred thousand chickens per day.[118]

The chicken processing companies also became vertically integrated into the companies that controlled chicken production rather quickly and thoroughly.[119] This vertical integration has become prevalent throughout the animal slaughtering and processing industry, although no sector has become as integrated as the chicken sector.[120]

Beyond just the chicken slaughtering and processing sector, the industry as a whole has been characterized by horizontal integration, which also contributed to employer ascendancy. Integration and consolidation became pronounced after World War II and increased quickly: by 1983 there were one-third fewer slaughterhouses than there had been in 1945.[121] Further, between the mid-1980s and the mid-1990s, another third of the meat processors in the United States had gone out of business because they could not compete with the large companies.[122] The power in the industry became progressively consolidated in the hands of a few companies, notably Tyson Foods, Pilgrim's, Sanderson Farms, and Perdue Farms, which controlled 69.28 percent of the broiler chicken industry as of 2012.[123]

The high degree of horizontal integration is reminiscent of what the industry looked like at the turn of the twentieth century, before the government had intervened and the consent decree had been signed. In fact, Horowitz concludes that the current oligopoly in the industry has returned slaughterhouses and workers to the conditions of the early twentieth century, which predated the gains made by unions.[124] In describing this resurgent consolidation, in the second half of the twentieth century, Aduddell and Cain warn, "how much further history will go in repeating itself remains to be seen."[125] Fewer companies has meant less competition and has given the companies in the industry the space and time to amass significant power and political clout relative to workers in the industry. The power of companies in meat production and processing is evidenced by the "acceptance of environmentally questionable development plans by state governments, the reluctance of federal regulators to require changes in processing methods that would reduce health hazards, [and] the setbacks to unions who sought to organize meat industry workers."[126]

It is not only that a few producers and slaughterhouse companies have become ascendant, there has also been significant consolidation in supermarkets and fast food chains, which has impacted producers and processors. Prior to the development of large supermarkets and "boxed beef" type of processing, consumers purchased their meat directly from relatively small butcher shops. These shops were dependent on the large meatpacking companies to supply the meat, and the large meatpacking companies generally had the power to set the terms of the deal. As supermarket chains gained access to consumers they cultivated more power in dealing with the meatpacking industry: the industry soon relied on supermarkets and fast food chains as mediaries between them and consumers. In fact, Horowitz argues that the supermarkets were responsible for the decline of the old dominant meatpacking companies in the second half of the twentieth century, which made it possible for newer companies, such as IBP, to gain a foothold in the industry. The amount of shares the old top four companies controlled declined from 52 percent in 1950 to 25 percent in 1972.[127] By the 1980s, many of these companies had gone out of business because they were unable to compete with IBP and other companies that had been able to undercut them. The meatpacking companies that remained had to compete to get contracts with the large supermarkets and fast food chains. They were under pressure to produce the product even more cheaply and uniformly. As a result,

> fighting for a limited number of lucrative contracts, meat firms were impelled to do whatever they could to keep their prices down. Going beyond the minimum requirements of government regulations thus was an unwise business decision, as allowing processing, labor, or supply costs to exceed competitors' would have damaging consequences.[128]

This development worked against the interests of workers across the industry.

In sum, the later period of industrialization in the industry has been characterized by a decline in unionization as a result of *economic restructuring, working-class fractionalization,* and *employer ascendancy.*[129] Workers lost some notable gains that had previously been made in the industry, and although the industry has become highly competitive, between a few companies in particular, it remains profitable. For instance, if one had purchased $10,000 worth of shares in Tyson in 1975, by 2005 they would have been worth $6.96 million.[130]

Using the techniques discussed in this chapter, the industry has been able to keep their production costs and the cost of their products down. Marcus demonstrates that the industry has been able to keep its costs relatively static by putting their costs in the larger context of the rising cost of living in the United States since 1950. He finds that, while the price of the average new home has increased 1,500 percent, the price of eggs and chicken meat, for instance, has not even doubled. This prompts him to ask, "why is the industry resisting inflationary pressures?"[131] His answer is that the industry has become more efficient. I would suggest qualifying this answer by specifying that the industry has become efficient at controlling their production costs, but not necessarily efficient in keeping externalized costs down.

THE GROWING DIVIDE BETWEEN PRODUCTION, PROCESSING, AND CONSUMPTION

Despite the relocation of slaughterhouses closer to livestock production in the most recent era of slaughterhouse development, there has nonetheless been an increasing divide between production and processing, on the one hand, and processing and consumption, on the other. In the first instance, animals are taken from the facilities that specialize in production ("farms") and transported to the facilities that specialize in slaughter and processing. Even though these facilities may be owned by the same company and are geographically closer than they used to be, there is still an important logistical distancing. Analysts would point to the efficiencies of this type of specialization, but importantly it also functions to make it easier for the producers to send animals off to slaughter because they do not encounter the realities of it.[132] Some of the farmers Wilkie interviewed for her research on livestock workers in Scotland reported that they could not slaughter animals and that they even disliked or looked down upon those who do. She reports, "One farming contact said that slaughterers are little more than 'animals,' which implied that producers are more human. That farmers can distance themselves from the killing stage provides a practical basis for making such a claim."[133]

The consumption of animal-derived products has also become further distanced from their processing during this era. Gouveia and Juska suggest two factors that have contributed to this separation.[134] First, slaughterhouse employees have become separated from the community, and from consumers by extension. As discussed earlier, prejudice used to be expressed against butchers for the work they did killing and processing animals.[135] Through the process of industrialization, the prejudice was extended to slaughterhouse workers—a much more diffuse group. This prejudice against these workers and their demographic backgrounds has been associated with xenophobia in some slaughterhouse communities. In short, many slaughterhouse workers have been alienated from the communities in which they live. Therefore, the limited connections that had existed between consumers and those who slaughter, process the animals, and cut the meat have been severed.

The second factor Gouveia and Juska point to that has furthered the separation between consumption, production, and processing is the social construction of the consumption of food as a very individual-level phenomenon, which finds broader support from the emphasis on individual choice within neoliberalism.[136] Thus, the ethical and logistical issues associated with production and processing are bracketed out in favor of one's ability to choose what s/he wants to eat. Consumption gets reduced to a lifestyle choice.

Of course, the notion of individual choice is somewhat an illusion in this context. Due to the high degree of corporate consolidation discussed earlier, consumers actually do not have many options regarding which companies to do business with. Further, it is difficult for individual consumers to gather information about corporate practices and the food supply more generally, which has actually created a market for books detailing the origins of what we eat.[137] Many consumers do not read such books, nor do they actively search out information on the production of their food, and they may assume that their interests and those of the animals used are being protected by the government in the form of regulations.

Regulation of the industry provides some visibility. However, the existence of regulations, as limited as they are, may further secure the divide between consumption, production, and processing. In particular, they give consumers the impression that through this visibility the best interests of all involved are being protected; thus, there is no need to look beyond one's own consumptive behavior. Yet these regulations are developed and applied within a larger context of power imbalances, which has important consequences. Pachirat argues that power operates in the context of the slaughterhouse not only through its concealment from the larger culture, but also through surveillance and the bridging of distance within its walls. He writes, "surveillance and concealment work together . . . quarantine is possible in, and perhaps even enabled by, conditions of total visibility."[138]

REGULATING ANIMAL SLAUGHTER

As mentioned earlier, Upton Sinclair's book *The Jungle* inspired the enactment of legislation regulating the meat industry in 1906: the Pure Food and Drug Act banned interstate trade in adulterated or mislabelled food and drug products, and the Federal Meat Inspection Act expanded inspection from hogs to all livestock being exported across state and international lines and further gave the USDA increased regulatory power.[139] Nonetheless, there continued to be problems in the industry.

In the 1960s, it was revealed that slaughterhouse companies were exploiting a loophole to evade federal inspection. By processing at least one quarter of their meat in "intrastate plants" and restricting shipment of the product to within the state, the companies were not technically engaging in interstate commerce and therefore avoided federal inspection. In response, the Wholesale Meat Act was passed in 1967. The act requires states to regulate intrastate commerce in the same manner that interstate commerce was. Of course, the act is premised on the assumption that federal regulation of interstate commerce is sufficient. The congressional hearings on the issue also exposed the commonality of slaughterhouses processing "4D livestock" (those dead, dying, diseased, and disabled).[140]

In the subsequent decade, animal slaughter regulations were updated through the Humane Methods of Slaughter Act, enacted in 1978. From this point on all federally inspected slaughterhouses had to comply with the act. Among other things, the act requires the stunning of cattle and pigs before slaughter.[141] Notably, however, the act does not apply to chickens and other birds or to animals killed as part of religious or ritual slaughter.[142]

These far-from-radical legislative changes were met with increased industry agitation against regulation. The industry argued that regulation of meat processing practices was unnecessary and costly. They instead suggested, in line with the escalating neoliberal ethos, leaving regulation to the market: let the consumer vote with their dollar and let competition shape practices in the industry.[143] There are, however, some significant potential problems with this approach. First, as discussed earlier, consumers often do not have enough information about how food is produced to make an educated decision when voting with their dollars. Second, even if the problem of informing the public could be overcome, those with limited resources to spend on food would still have their choices curtailed.

The traction of industry complaints about regulation was limited somewhat by a number of meat contamination outbreaks, beginning in the 1980s. Publicized cases of meat contamination—particularly the Jack in the Box case, where an E. coli 0157:H7 outbreak cost four children their lives and sickened hundreds of other adults and children—increased the public's concern about the meat supply. These cases began to publicly bridge the distance between consumption on the one hand and production and processing on the other. In short, they

> revealed the volatility of artificially-constructed production-consumption distance, expressed as well as constituted by regulatory regimes premised on individualistic constructions of consumers. In a very short period of time, these very same consumers turned to the government demanding collective freedom from contaminated meat.[144]

The government responded to the public outcry, and in 1994 the USDA categorized E. coli 0157:H7 as an adulterant that should be tested for. The slaughterhouse industry challenged this designation and sued the USDA in an attempt to prevent it from conducting bacteriological testing on meat. The legal ruling was eventually in favor of the USDA; as a result, the Pathogen Reduction Act came into effect in 1996.[145]

The act introduced bacteriological meat safety regulations and seemed like a win for food safety advocates; however, it had three significant limitations. First of all, any regulation is only as good as its enforcement, and enforcement of the act was limited because the USDA had its power to conduct inspections reduced. Instead of conducting frontline inspections, the agency was charged with verifying records created by the companies. The companies would essentially be policing themselves. Second, the act was not designed to get the industry to eliminate the origins of contamination; instead the industry could develop new technologies to *de*contaminate meat, such as irradiation, carcass rinsing, and ozonation. In essence, "all of these technologies are directed at taming, or fixing 'nature' rather than 'fixing the system.'"[146] Finally, the act had the perhaps unintended consequence of augmenting the already significant consolidation in the industry. Capital is required to implement the decontamination technologies, and the amount needed represents a much more significant allocation of resources for smaller companies.[147] As a result, some smaller companies have not been able to comply.

Limitations of this sort have prompted serious concern among various parties regarding the regulations governing animal slaughter and meat processing. Animal rights and welfare advocates, as well as food safety advocates, have been particularly vocal with their concerns. What is perhaps more surprising is that criticisms of these regulations have also been articulated by those charged with enforcing them. For instance, twenty-four USDA regulators submitted a letter to the National Academy of Sciences, wherein they state, "In good conscience, we can no longer say that we know USDA approved beef is wholesome."[148] In spite of the public airing of these concerns, the regulatory environment remains unchanged.

In response to the continued limitations of government regulations and the strength of the industry in resisting them, a new tactic has recently been employed in an attempt to improve food safety and animal welfare: targeting large fast food companies that have contracts with the meat production and processing companies. Animal advocates first targeted McDonald's, and after significant pressure, in 2000 the company agreed to undertake surprise welfare inspections of their meat suppliers. The threat of these inspections gives the slaughterhouse companies that want to retain lucrative McDonald's contracts incentive to clean up their act. Some other fast food chains have followed McDonald's lead. While these surprise corporate inspections have had positive impacts on the treatment of animals and presumably food safety in some slaughterhouses, they also highlight the limitations of regulation by the USDA.[149]

THE SLAUGHTERING PROCESS

Up to this point, the slaughter of animals has been discussed in fairly abstract terms. I would like to conclude this chapter by providing a snapshot of what industrial animal slaughter in the most recent era in slaughterhouse development looks like, as most readers will not have firsthand knowledge of the process. Additionally, consumers certainly do not receive detailed information on the transformation of animals into meat from the producers, meatpacking companies, retailers, or the government: information about the bodily transformations is intentionally absent. Utilizing cultural theory, Shukin has analyzed how power is exercised through the bodies of animals in her book *Animal Capital: Rendering Life in Biopolitical Times*. She cites Michel Foucault, who wrote that "control of society over individuals is not conducted only through consciousness or ideology, but also in the body and with the body."[150] The animal slaughtering process, particularly in its most recent iteration, is perhaps the best illustration of how the control of a society can manifest itself in the bodily, as described in what follows.

The slaughter process varies by species of animal. A captive bolt gun is used with cattle and horses to stun them prior to being killed. They are then "stuck" (cut) in the neck or chest so that they bleed out (see figure 15).[151] Pigs can be stunned with gas (carbon dioxide), as can poultry (with carbon dioxide, argon, and nitrogen); however, they are most commonly electrically stunned.[152] This stun wears off quickly. Gregory therefore suggests sticking pigs within twenty-three seconds of their stunning, which is when they can begin gaining consciousness. This certainly does not give the "sticker" much time to do his job. One solution would be to use higher voltages for the stunning of pigs and chickens to lengthen the time between stunning and regaining consciousness. In fact, European countries use higher voltages in electrified stunning than required in the United States.[153]

Forms of religious slaughter, such as shechita (Judaism), halal (Islam), and Jhatka (Sikh) do not involve stunning (although some countries, such as New Zealand and Australia require

Figure 15. Cow slaughter. (Source: People for the Ethical Treatment of Animals. Used with permission.)

stunning in certain cases). Shechita slaughter is supposed to involve slicing the animal's neck in one movement. Jhatka requires decapitation via one stroke of the cutting instrument. The halal method varies somewhat depending on the custom and experience of the slaughterer and can involve decapitation or severing the spinal cord.[154]

In industrial animal slaughter time is not wasted, and immediately after the animal has his/her throat slit to bleed out, s/he is transferred along the disassembly line to be dismembered. Typically, after being stunned, cattle are placed on a conveyor belt and transported to a shackler who attaches a chain to a hind leg so that the animal can be raised up and transported past an indexer, who attempts to space the animals evenly apart and looks for signs of consciousness. If signs of consciousness are observed the indexer uses a captive-bolt handgun to shoot the cow's head again. The disassembly line then takes the cow to another location to have his/her throat slit. The goal is to sever the carotid arteries and jugular veins. The line then takes the animal to the next area for electrocution, which stimulates the heart to expel the rest of the blood. Ideally, the cow is dead by the time s/he reaches the next stage—the tail ripper—although this is not always the case. The body is then dismembered further in stages for processing and packaging (see figures 16 and 17).[155]

Chickens are small enough and uniform enough to be decapitated by a machine after they are stunned. Their bodies then go through scalding water and then a defeathering machine. They then go through the evisceration process, done largely by hand and mostly by women (see figure 18).[156]

After stunning, pigs are placed on the line and have their carotid and jugular veins cut to drain the bodies of blood. They are then scalded, dehaired, and "debunged" whereby the rectum is removed to mitigate contamination through the remainder of the process. An automatic saw is then used to eviscerate the pigs and remove the head and spinal cord. The weighing and trimming work then begins (see figures 19 and 20).[157]

The speed of the disassembly line, particularly in the United States, is extremely fast. For instance, for cattle slaughter the common line speed is currently four hundred animals an hour. To put this into perspective, a fast pace in the United States in the 1970s was 170 an hour and in Europe it is currently 60 per hour. The current rate means that a new cow is stunned every nine seconds.[158] One consequence of this extremely fast line speed is that it increases the chance of worker injuries.[159]

Another consequence of accelerating line speeds has received growing attention in recent years: it increases the number of animals who go down the disassembly line while still conscious. In 2001, the *Washington Post* published an article titled "'They Die Piece by Piece': In Overtaxed Plants, Humane Treatment of Cattle Is Often a Battle Lost," wherein they detail the disturbing frequency of cattle being improperly stunned and dismembered while conscious.[160] The reporter interviewed a man referred to as Moreno, who had worked for twenty years as a "second-legger" in an IBP slaughterhouse. The reporter explains,

> The cattle were supposed to be dead before they got to Moreno. But too often they weren't. "They blink. They make noises," he said softly. "The head moves, the eyes are wide and looking around." Still Moreno would cut. On bad days, dozens of animals reached his station clearly alive and conscious. Some would survive as far as the tail cutter, the belly ripper, the hide puller. "They die," said Moreno, "piece by piece."[161]

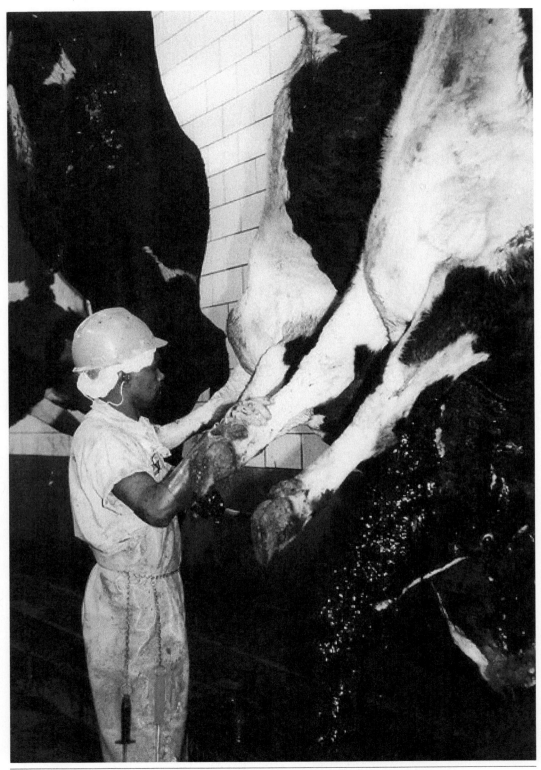

Figure 16. The cattle disassembly line I. (Source: ©Humane Farming Association. Used with permission.)

Figure 17. The cattle disassembly line II. (Source: ©Humane Farming Association. Used with permission.)

Figure 18. Chicken disassembly line. (Source: L. Parascandola, courtesy of United Poultry Concerns. Used with permission.)

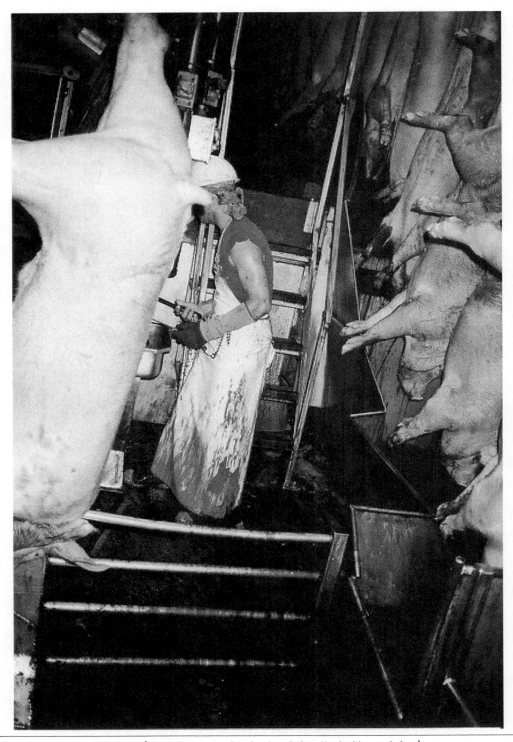

Figure 19. Pig disassembly line I. (Source: ©Humane Farming Association. Used with permission.)

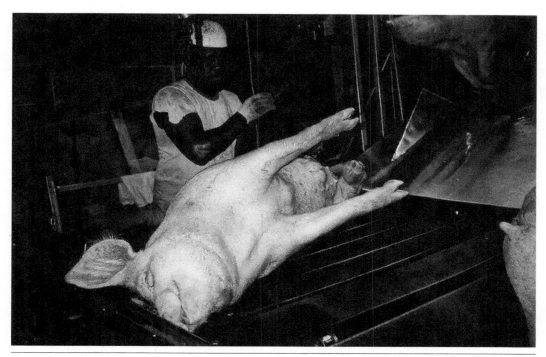

Figure 20. Pig disassembly line II. (Source: ©Humane Farming Association. Used with permission.)

A veterinarian who used to be the chief government inspector at a slaughterhouse in Pennsylvania corroborated Moreno's story, stating that the dismemberment of live, conscious animals occurs in industrialized slaughterhouses daily.[162]

The reporter also explains in the article that enforcement of welfare laws inside slaughterhouses by the USDA is weak. Even when a plant is cited for violations, such as a slaughterhouse in Texas that was cited twenty-two times in one year for violations including cutting hooves off of live cattle, no further action is taken. Instead of monitoring these problems more closely, beginning in 1998 the USDA stopped tracking the number of humane slaughter violations their inspectors uncover in slaughterhouses every year. As mentioned earlier, some inspectors have become so frustrated that they have taken to complaining publicly.[163]

Academic studies have corroborated the commonality of animals being improperly stunned for slaughter. A recent ethnography of work inside a cattle slaughterhouse by Pachirat provides evidence that the improper stunning of cows is common enough that, at least in the slaughterhouse he worked in, special gates have been installed along the disassembly line to prevent conscious cows from running any further down the line.[164] The dismemberment of conscious animals inside slaughterhouses was also corroborated by Gail Eisnitz, chief investigator with the Humane Farming Association, through her interviews with slaughterhouse workers, who told her it happens with shocking frequency. The significance of this reality for the workers is illustrated by the following statement by one of Eisnitz's informants:

> Every sticker I know carries a gun, and every one of them would shoot you. Most stickers I know have been arrested for assault. A lot of them have problems with alcohol. They have to drink, they have no other way of *dealing with killing live, kicking animals all day long*. If you stop and think about it, you're killing several thousand beings a day.[165]

These are the realities that consumers are spared from because a whole workforce is employed to keep meat processing and consumption separate.

The historic transition of slaughterhouses from the public's gaze drove a wedge between livestock production and processing, on the one hand, and between livestock slaughter and consumption, on the other. Consumers no longer saw where the meat they were eating came from, and farmers and breeders lost view of the destination of their animals.[166] Today this separation is even more pronounced. There are now fewer, larger slaughterhouses that kill so many animals per day that it nearly defies comprehension. The buildings are nondescript and secured. Very few see within their walls, and the number who do will likely decline even further due to "ag-gag" legislation that criminalizes taking film footage and photos inside the meat production and processing plants. Even the end product of this entire process—meat—carries with it few reminders of the animal(s) it once was.

Consuming Animals as Food

THUS FAR WE HAVE EXPLORED THE PREAGRICULTURAL RELATIONS WITH ANIMALS, THE subsequent industrial production of animals for human consumption, and the processing of their bodies to turn them into consumables. In recent history the production of livestock animals and the processing of their bodies have been separated from each other in important ways. They have also been separated from the process of consumption—the focus of this chapter—and all three processes have been conceptually distanced from their consequences.

The separation of consumption from production and processing has been particularly dramatic. As a result, "we see consumption as an end in itself, and we do not consider what have been the means to that end."[1] Of course, this is not unique to animal-derived products; the consumption of many products overshadows the means of bringing those items to market. For instance, consumers generally do not think about the garment workers producing their reduced-cost clothing. So it is not that surprising that as a culture we tend not to envision the animals whose bodies we dine on or the workers who toil in precarious working conditions to transform animal bodies into edible parts.

While the previous chapters have attempted to bring into view the means (i.e., production and processing) to the end, the overarching goal of this chapter is to critically examine the end itself: meat consumption. In western cultures, current patterns and practices of meat consumption have been normalized.[2] This chapter examines how consumption has varied over time and place, which makes it possible to begin unpacking this normalization. In doing so, this chapter further contributes to bridging the distance between consumption and its antecedents (production and processing) and their consequences.

While the constructed gaps between production, processing, consumption, and their consequences widened throughout the twentieth century, recent years have witnessed some countervailing forces that have begun to narrow the gulf between these elements. One such force has been popularized literary and cinematic works. Books by Jonathan Safran Foer and Michael Pollan, both critically acclaimed authors, have spurred some members of the general public to think more critically about the animal products they are consuming.[3] And documentaries such as *Food Inc.* and *Forks over Knives* have provided another medium for bridging the chasm and questioning consumptive practices.

Activities within the industry have also recently unwittingly helped to forge connections for the general public between production, processing, and consumption. Gouveia and Juska identify two junctures within the industry that in particular "threaten to over-expose the problematic linkages between industrialized meat production and meat consumption."[4] The first is food-borne illness, which can bring the animals people consume into their consciousness. In the wake of cases that receive media attention, at least some consumers begin to contemplate how meat becomes contaminated, prompting questions such as what does the process of producing meat look like and what are its flaws? This can be a significant entrée into thinking more critically about meat production, processing, and consumption. The second juncture where connections between the production and consumption of meat come to the fore is via workers in the industry. Workers have received attention of late in the context of allegations that the industry is using undocumented immigrant labor. The public generally does not think much about workers in the industry, but the vitriol around the topic of undocumented immigrants has drawn attention to this group of workers. Slaughterhouse and animal production workers also occasionally come to the attention of the public when workplace injuries and illnesses receive attention, such as cases where workers are overcome by manure vapors and die. The connection between production, processing, and consumption vis-à-vis workers is particularly salient in communities where large production and processing facilities operate.

When connections between the animal agriculture industry and consumption are made in these ways, governmental and corporate responses commonly attempt to reestablish the divisions between production, processing, and consumption. For instance, during food-borne illness outbreaks, the government and corporations have suggested that increased regulation of production and processing is not the solution and instead that individual consumers are knowledgeable enough to consume the safest products and cook them to protect against illness, reinforcing the neoliberal veneration of market forces and individual choice.[5]

The tension between maintaining the distance between production, processing, consumption, and their consequences, on the one hand, and the threats posed by points of convergence (e.g., via books and films, incidents of food-borne contamination, and issues with workers in the industry), on the other, are elucidated and woven throughout this chapter. The chapter begins with an examination of the history of meat consumption, followed by an examination of current consumption patterns, how meat products are marketed to consumers, attitudes toward consumption, and how consumptive practices are shaped by the species of animal in question.

This chapter intentionally follows the preceding two on production and processing to avoid giving the impression that consumptive desires necessarily precede and drive the industry. For, as Dauvergne rightfully points out in his examination of the environmental consequences of consumption, "the notion that demand arises from the innate needs and cravings of freethinking consumers—even for a basic need like food—is far too simplified."[6] Demand is shaped by various external forces, such as cultural factors, governmental policies, and industry interests, and not simply an individual desire for animal protein. This chapter provides insight into these complicated and multifaceted influences on the use of animals as food.

CHANGING CONSUMPTION RATES, PRACTICES, AND PERCEPTIONS

Western cultures have developed what Franklin perhaps conservatively refers to as a "special fondness for meat."[7] What is it about meat that makes people so fond of it? It is not just the taste that endears many people to it; there are other factors that influence consumption and have made meat a significant part of western diets. These factors can be grouped into three general categories (which are neither mutually exclusive nor exhaustive): cultural, industry, and governmental. These factors are discussed throughout this chapter; however, a brief description of each at the outset is warranted.

At the cultural level, meat is important for two main reasons: it has developed what anthropologists refer to as communal significance, and it has come to represent power, particularly masculine power. The communal nature of meat can be traced back in history to the practice of subsistence hunting and small-scale agriculture where the slaughter of large animals for consumption produced an abundance of meat and the excess was shared with others, making it "the basis for social meals or gifts and . . . associated, therefore, with . . . social relations."[8] Even though most people no longer kill their own animals for meat, it has retained this cultural significance to varying degrees. Culturally, meat has also been associated with power and masculinity, and although one might think that this connection would be weakening as conceptualizations of masculinity have multiplied and taken on different forms recently, there is evidence that the connection between meat and masculinity has been strengthened in the process of a backlash against changing constructions of gender.[9]

Industry influences on consumption have also been significant. For several decades now the animal-industrial complex has been marketing meat as an essential part of the human diet and promoted ever higher levels of consumption.[10] The industry has also shaped what types and cuts of meat people consume. By keeping the costs of meat artificially low, through externalizing some costs and securing subsidies to cover others, the industry is also able to influence consumption. The industry not only influences consumers in these rather direct ways, it also does so in a more indirect way through lobbying the government, which in turn influences consumption through specific policies.

As discussed in previous chapters, governmental policies have facilitated increases in production in the industry, which has made products more widely available for consumers and, importantly, relatively cheap. The government has also influenced consumption more directly by providing nutritional guidelines that promote meat consumption, particularly as a source of protein.[11] Although it may be difficult to envision from this specific vantage point in time, governmental pressures toward increased consumption are relatively recent. It was not long ago that some western governments were restricting meat consumption among the general public.

Consumption Pre–Twentieth Century

The fact that meat has become the focal point of the western diet is undeniable, although it is not clear exactly when it took the center stage it occupies today. Historically, there were actually controls placed on meat consumption in some locations. For instance, in medieval Britain meat consumption was controlled by the state and the church. These controls were at least partially motivated by the church's desire to have its adherents practice self-control. It was feared that eating meat might make people more animalistic and thus less likely to be able to control their passions.[12]

For the next several centuries, meat consumption remained relatively low in Europe and North America by today's standards. The settlers in the North American colonies had such a small amount of meat available that it was difficult for them to ration it to last through the winter. Beginning in the eighteenth century, increased supplies of salt and livestock made the rationing progressively easier.[13] In the subsequent century, meat still made up less than 15 percent of the protein Europeans consumed, although consumption increased later in the late nineteenth century as they began importing meat from the United States and Australia.[14] Increasing supplies and urbanization steadily facilitated increases in meat consumption in Europe and North America from the late nineteenth century onward.[15]

It is difficult to ascertain exactly how much meat individuals were consuming prior to the twentieth century because consumption records were not kept. Widow allowances and slave rations in the United States, however, provide some insight into levels of consumption. Widow allowances (rations provided to women after their husband had died) reportedly contained 120 pounds of meat in the early eighteenth century and 200 pounds by the early nineteenth century. Slaves in the South were allocated on average 150 pounds of meat each. Horowitz suggests we can extrapolate from these statistics that the average per capita annual consumption of meat was 150–200 pounds in the nineteenth century.[16]

Although these rationing statistics provide us with general information on overall consumption, they do not permit a nuanced understanding of the types of meat being consumed at this time and the sociodemographic variables that impacted consumption. Other sources indicate that at this time in the United States, pork was the most commonly consumed meat because it could be cured and consumed year round. Pork reigned supreme until beef was made more accessible through the development of refrigeration technology and the centralization of markets facilitated by urbanization. Growing incomes also made it possible for more people to consume this type of meat and increased the demand for beef across geographic regions and racial and ethnic groups. It became the most popular meat in the United States by the twentieth century.[17] Beef would retain that title for much of the twentieth century.

Consumption in the Twentieth Century

Consumption of meat in the twentieth century fluctuated against the backdrop of wars and economic crises. Observers have commented that, in spite of those fluctuations, consumption became increasingly democratized throughout the century: increasing numbers of people were consuming increasing shares of meat. Growing industrial production and rising wages translated into the greater affordability of meat, and improved transportation meant that consumption was no longer tethered to geography. This democratization, however, brought with it increased consumption rates and resultant concern in the latter part of the century about the healthfulness of consuming high-fat meats.[18]

Consumption studies became more common in the twentieth century, creating better data about consumption for this period than for earlier ones.[19] According to a 1909 study, the average person in households with incomes less than $1,000 per year was consuming 136 pounds of meat annually and those in households with incomes greater than $2,000 were consuming just over 200 pounds on average per year. Not only was the amount of meat consumed still impacted by income early in the century, but so was the type of meat consumed: those with higher incomes consumed greater amounts of beef and poultry.[20]

The steadily climbing rate of meat consumption observed at the beginning of the century was derailed by a crash in the market in the 1920s caused by overproduction in the industry after World War I. The crash took a serious toll on farmers and ranchers, and approximately one seventh lost their farms.[21] Meat consumption further declined in the following decade during the Great Depression. Records indicate that during the mid-1930s, per person consumption dropped to between 85 and 122 pounds per year on average, depending upon geographic region and race.[22]

Demand for meat did not increase again until World War II. During that war, the industry could not keep up with the demand, and rationing was implemented.[23] Despite the increased demand, meat consumption still had not returned to the level it had been at the turn of the twentieth century, likely due to rationing by the government. (See the rationing poster produced by the U.S. Food Administration in figure 21.) Consumption rates, however, were about to make a resurgence, and even surpass previous levels. The end of the war and meat rationing contributed to significant increases in consumption in the 1950s and 1960s, which ushered in higher levels of meat consumption per capita on average than ever before recorded.[24]

Increasingly income became less of a determinant of who could eat meat.[25] By the mid-1960s, two-thirds of U.S. families could afford to consume the ultimate decadent meat product: steaks.[26] Meat, particularly the more expensive cuts, became a symbol of the growing prosperity in the United States and elsewhere. The meat industry itself also became symbolic of progress; "The modernization of meat production in the West over the course of the twentieth century was a key marker of economic and social progress because it made a former 'positional' good available to everyone."[27] Meat became an important part of Americana, materially and symbolically.

As meat consumption increased throughout the twentieth century, the types of meats consumed also changed. Although consumption of chicken is now ubiquitous, up until World War II it was actually considered a luxury food. Cost had earned it this reputation: at mid-century it cost approximately the same as round steak. A progressive drop in price contributed to its increasing popularity in the second half of the century. Other factors also contributed to this growing popularity, including that it had not been rationed during the war like other meats, therefore people had become more familiar with its use and preparations; it became more widely used in the growing fast food industry; consumers saw it as a healthier alternative to high-fat meats; and the chicken industry successfully marketed their product as a versatile type of meat (up to this point it was not actually considered meat). As a result, poultry became as popular as beef by the 1980s.[28]

By the late 1990s, poultry consumption was consistently exceeding beef and pork consumption, and the gap has grown since. The highest level of poultry consumption was recorded at 74.2 pounds in 2006. Beef consumption peaked at 91.5 pounds per person on average in the mid-1970s. Pork consumption remained fairly stable throughout the century; the highest level of consumption was 53.6 pounds, recorded in 1944. As of 2011, the average person consumed 55.5 pounds of beef, 43.2 pounds of pork, and 71.9 pounds of poultry per year.[29]

The cuts of beef, pork, and chicken consumers were eating also changed throughout the century, although consumers were not necessarily the ones driving these changes. As production became increasingly industrialized, the industry standardized the type of cuts made, and those became the cuts that consumers became accustomed to. Supply-driven consumption was not new to the industry. The nineteenth century had witnessed changes in beef and pork production that consumers were induced to adapt to. At first, consumers were wary about eating meat from cows that had been killed days prior and shipped to market. Yet the industry went ahead with this form of processing, and the reduced cost influenced consumers to give it a try. Within a few decades this type of beef consumption was commonplace.[30] Likewise, in the twentieth century, consumers became accustomed to standardized cuts, such as ham and bacon.

Figure 21. First World War—Eat more corn, oats and rye products. . . . Eat less wheat, meat, propaganda poster, illustration by L. N. Britton, 1917. (Source: ©De Agostini/The British Library Board, 1294549. Used with permission.)

Several pork companies started producing bacon and ham cuts, branding them with their company logo and marketing them to consumers (see figure 22). They used these processed foods as a vehicle to get their company name out to consumers, with the goal of developing brand loyalty. Consumers bought it, and in the twentieth century could not get enough of it. Although consumers were no doubt creating a demand for these more convenient cuts, the industry was promoting them because they could add value to and charge more for these products. Changes in chicken consumption over time illustrate the convenience food trend. In the 1960s, 83 percent of the chicken consumed was in the form of whole chickens. By the mid-1990s, 86 percent of chicken sales were for chicken that had been processed, such as in the form of the popular chicken nugget.[31]

In addition to changes in the types of meat consumed, the dramatic growth in per capita consumption of all types of meat over the course of the twentieth century is also noteworthy. The highest level of recorded consumption was 184.1 pounds in 2004. The lowest annual consumption was recorded in 1935 (during the Depression) at 82.8 pounds.[32] Only recently, in 2008, has the upward trend in per capita meat consumption actually been reversed, although it is unclear if this is just a temporary blip or part of a new trend. The most recent data (for 2011) indicates that the average person in the United States consumes 170.6 pounds of beef, pork, and poultry annually.[33] To put this into perspective, meat consumption has increased by approximately 54 percent since 1950.

Symbolism: It's What's for Dinner

The industrialization of meat production and processing and the resultant growth in the meat supply certainly made increases in consumption possible; however, the linking of meat with prosperity and power made it highly desirable. In the twentieth century, the beef industry mounted a ubiquitous advertising campaign with the tag line, "Beef: It's what's for dinner." But they were not just marketing food: they were also marketing a lifestyle and a way of "doing gender."[34] Post World War II, the American dream became a reality for many people: having a job that paid a decent wage, getting married, owning a home, and owning a car. The dream, however, was seemingly not complete without burgers and steaks on the barbeque in the white-picket-fence-encircled yard, as well as trips in the car to the local fast food joint for a burger and fries,[35] and a large turkey on the table for Thanksgiving (illustrated by the postcard pictured in figure 23). More people wanted access to this dream, and increasingly more people could afford it. The industrialization of production and processing, along with handsome government subsidies, made meat economical, if not cheap. In fact, adjusting for inflation, meat became increasingly inexpensive over the years, reaching its fifty-year lowest price in the 1990s.[36]

The fusion of meat and prosperity is perhaps only rivaled in its symbolic power by the linkage of meat with power and vitality. This linkage is said to stem from the belief

> that the strength and vitality of animals can be assimilated into the human body through digestion. Meat is like no other food, because only meat is like humans corporeally. Because animals are made of similar material, the consumption of their flesh is the clearest, most direct mode of maintaining the human body. It is significant perhaps that the most highly valued meats derive from animals whose bodies are both bigger and stronger than the human body.[37]

Figure 22. Advertisement for Armour smoked ham, 1948 (color litho), American School (20th century). (Source: Private Collection/Peter Newark American Pictures/Bridgeman Images. Used with permission.)

Figure 23. Give Us This Day Our Turkey Postcard, ca. 1908. (Source: Lake County Discovery Museum/UIG/ Bridgeman Images. Used with permission.)

Variations of this belief persist today. There remains a strong association between meat consumption and masculinity, particularly vis-à-vis physical strength.[38] This is not unique to the United States: meat is associated with masculinity in many countries.[39] This association, however, has historically been quite pronounced in the United States and has exhibited a resurgence in recent years as socioeconomic changes have destabilized traditional conceptualizations of masculinity.[40]

Recent empirical research provides evidence of this association between meat consumption and hegemonic masculinity. An analysis of televised cooking shows finds that farm animals and concern for them are feminized, whereas the slaughter of these animals and meat consumption are celebrated and positively associated with masculinity.[41] Another study examining television advertisements for meat products finds that vegetarianism is cast as a threat to masculinity, which advertisers are quick to exploit by depicting meat consumption as reestablishing the gendered status quo.[42]

Although gendered norms are changing, and there are certainly more male vegetarians and vegans than in the past, research indicates that men still consume on average 74 grams more meat per day than women.[43] Although differences in purported nutritional needs are no doubt related to this gendered difference, it may also be at least partially related to reaffirming masculinity and gendered differences in concern for animals. For years studies have documented significantly higher levels of concern about animal well-being among women than among men; however, the role of mediating variables in this relationship was unclear. For instance, these studies could not exclude the possibility that women on average care more about animals because they generally bear more responsibility for caring for children and the household, of which animals are often a part. A recent study of Ohio residents, however, statistically controlled for a number of potentially mediating variables, such as household responsibilities, presence of children, and various experiences, and found that women are significantly more likely to be concerned about animal well-being, "pointing to the overriding importance of gender as a social structural location."[44] Thus, gender and meat consumption may be related on the one hand because the consumption of meat may be a vehicle for reaffirming masculinity, and on the other because of greater concern for animal well-being among women.

There appear to also be gendered differences in how men and women explain or justify their consumption of meat. Psychologist Hank Rothgerber conducted a study with male and female undergraduate students and found that the male participants use direct strategies to rationalize consumption, such as denying that animals suffer, subscribing to beliefs in human superiority over animals, religious justifications, and claiming health benefits of meat consumption. Female students, on the other hand, tended not to employ direct justifications and instead use indirect strategies, such as dissociating animals from meat and avoiding contemplating the treatment of animals. He also found that higher rates of meat consumption among the male participants were significantly related to self-reported masculinity measures. Rothgerber observes that attempts to educate the public about the treatment of food animals and the benefits of vegetarianism have overlooked the possibility that men may receive and be differentially receptive to these messages; in short, such educational initiatives "may ignore a primary reason why men eat meat: it makes them feel like real men."[45] Even among vegetarians in western countries there appear to be gendered differences in motivation, with male vegetarians less likely to report being concerned about animal welfare and the environment than female vegetarians.[46] In sum, gender has continued to be a salient factor in consumptive behavior over time.

The New Century of Consumption

The dawn of the twenty-first century did not bring with it significant changes in meat consumption, although it was accompanied by a notable undercurrent of growing reflexiveness among some consumers. Horowitz observes that "America remained firmly, if somewhat guiltily, a meat-eating nation. . . . Meat remains a sign of the good life, the American life, and a valued item in our diets, even as Americans remain skittish about the wholesomeness of our food system."[47] Yet, by the end of the first decade of the twenty-first century, a measurable change had occurred: average per capita meat consumption began to drop off. As mentioned earlier, it remains to be seen if this is the beginning of a larger trend toward lower per capita meat consumption. What is evident thus far is that the decline in meat consumption has not been distributed evenly across the population.[48]

Recall that in the early twentieth century meat was a status symbol that not everyone could afford. Beginning in the mid-twentieth century, it became more accessible, and social class became less of a determinant of consumption. More recently, however, social class reemerged as an influence on meat consumption: now higher socioeconomic status (SES) has become associated with decreased meat consumption relative to other SES groups.[49] Occupation is also related to meat consumption: those in service or professional occupations consume less meat on average.[50] Race and ethnicity also influence total meat consumption and the types of meat consumed. Research indicates that Hispanics consume more beef than non-Hispanics, and Blacks and Asians eat more meat overall than Whites. The finding that Asians consume more meat than Whites in the United States may at first appear counterintuitive because meat consumption rates are relatively low in traditional Asian cultures. The researchers rightfully suggest this finding points to the importance of social context. Residing in the United States may contribute to increased meat consumption, perhaps at least partially because meat represents status among groups that are marginalized in the United States.[51]

These findings, combined with the changes in meat consumption over time discussed thus far in the chapter, indicate "that social structural factors clearly influence meat consumption habits. Therefore, meat consumption is clearly not the outcome of biological necessity, but a practice embedded within a complex of social forces."[52] One biological force that must be taken into account in discussing meat consumption, however, is the fact that the global human population is increasing exponentially, and this will invariably impact meat consumption.

THE IMPACTS OF POPULATION GROWTH: PAST, CURRENT, AND FUTURE

The average per capita consumption statistics discussed thus far only tell part of the story. Throughout the twentieth century as the meat consumption of individual citizens was increasing, so was the population of the United States. Thus, not only were individuals eating more meat, but there were more individuals to eat it here and around the world. The lowest gross amount of beef, pork, and poultry consumed was 9.5 billion pounds in 1915 (during World War I). Gross consumption increased fairly steadily from that point forward. Peak consumption occurred in 2007, with 55.4 billion pounds of beef, pork, and poultry consumed.[53]

Population size and consumptive practices have combined to make the United States the second highest meat consuming country in the world, with approximately 37.1 million tons of meat consumed in 2009. It is second only to China, which consumed 79.6 million tons of meat in the same year. Brazil comes in a distant third, having consumed 16.5 million tons in 2009.[54] We can expect that both per capita and gross meat consumption will increase in both Brazil and China as they continue down the development path charted by the United States.

Globally meat consumption rates are increasing the fastest in developing countries, and this trend is expected to continue.[55] Increases in per capita consumption in developing countries is approximately three times that in already developed countries.[56] Table 2 details the relative increases in meat consumption between 1980 and 2005 in developed and developing countries and regions. Not only are per capita meat consumption and GDP positively related, as reflected in increased individual consumption in developing countries, but these increases are amplified by the fact that many of the countries with increasing GDPs also have rapidly increasing populations.

The dramatic increase in meat consumption globally has necessitated a dramatic increase in the number of livestock animals raised and killed for meat, or what has been described in the literature as the *livestock revolution*.[57] The drivers of this revolution—human population, increasing incomes, and urbanization—are expected to continue to expand in most developing countries in the foreseeable future, thus amplifying consumption and the livestock revolution by extension. The projected increases are presented in table 3, which compares per capita meat consumption in the year 2000 and the projections for 2050. With the exception of North America and Europe, where consumption rates have nearly maxed out, the projected increases are significant.

Table 2. Per Capita Consumption of Meat by Region/Country Group/Country, 1980 and 2005(average kg per capita per year)

Region/Country Group/Country	1980	2005
DEVELOPED COUNTRIES	76.3	82.1
Former centrally planned economies (e.g., Russia)	63.1	51.5
Other developed countries (e.g., the U.S.)	82.4	95.8
DEVELOPING COUNTRIES	**14.1**	**30.9**
East and Southeast Asia	12.8	48.2
China	13.7	59.5
Rest of East and Southeast Asia	10.7	24.1
Latin America and the Caribbean	41.1	61.9
Brazil	41.0	80.8
Rest of Latin American and the Caribbean	41.1	52.4
South Asia	4.2	5.8
India	3.7	5.1
Rest of South Asia	5.7	8.0
Near East and North Africa	17.9	27.3
Sub-Saharan Africa	14.4	13.3
WORLD	30.0	41.2

Source: FAO, "State of Food and Agriculture."

Although the impacts of meat production, processing, and consumption are addressed more generally later in the book, there are three impacts related specifically to the increases of production, processing, and consumption in the developing world that I would like to address here to temper the temptation to frame this as a positive and unproblematic development. First, these increases are accompanied by a significant need for animal feed. The Food and Agriculture Organization of the United Nations estimates that demand for feed will increase by 1.2 percent each year.[58] Developing countries will therefore have to devote more of their subsistence cropland to producing animal feed (thereby reducing the crops for human consumption), create more cropland (contributing to deforestation and desertification), import animal feed from other countries, or amplify production on existing land through the use of chemical technology and biotechnology. Each option has its own associated drawbacks.

Second, the increasing demand for meat means that more animals are being raised and killed. This negatively impacts animal welfare, particularly as livestock production becomes industrialized in these countries to meet demand and profit goals. For instance, the dramatic increase in chicken consumption in developed and developing countries has had negative animal welfare effects. Chickens produce relatively little meat, which means that more of them must be raised and killed to satisfy production goals. Chicken production has been at the forefront of industrialization, and the conditions chickens endure are arguably the worst.[59]

Finally, although it is often assumed that increasing meat production and consumption in developing countries will mitigate malnutrition and poverty, this is not necessarily the case. Production in these countries is headed down the industrialization path that the United States

Table 3. Average Per Capita Kilograms of Meat Consumption by Geographic Region

Region	2000	2050 (projected)
North America and Europe	83	89
Central and West Asia and North Africa	20	33
East and South Asia and the Pacific	28	51
Latin America and the Caribbean	58	77
Sub-Saharan Africa	11	22

Source: FAO, "State of Food and Agriculture."

blazed in the twentieth century. There is no reason to think that the resultant concentration of production in the hands of the few in these countries will not have the same consequences there as it had in the United States: The masses will become dependent on corporations for their meat. Subsistence farming will be replaced by corporate farming, with devastating consequences. Therefore, instead of bringing health and prosperity to developing countries, "the alternative might be that the poor are driven out by industrial livestock producers; the one growing market the poor presently supply will be closed to them."[60]

Although the focus of this book is on the United States, it is important to keep this larger global context in mind. As population, per capita meat consumption, and industrialization increase in developing countries they will move closer to the current situation in the United States. Just as U.S. practices of meat production, processing, and consumption are being globalized, so will the impacts of exponentially increasing production, processing, and consumption in developing countries.

THE GROWING DIVIDE BETWEEN CONSUMPTION, PRODUCTION, AND PROCESSING: CONSUMPTIVE PRACTICES AND ATTITUDES

Levels of gross and per capita consumption are not the only things that have changed about meat consumption over time. Practices of and attitudes toward meat consumption have also shifted. Our hunter-gatherer ancestors had interdependent relationships with the animals they consumed. They viewed the animals as sentient beings and took killing them for human consumption very seriously; "In the West this relationship with food animals has been gradually eroded to the point where meat is separable from the animal it comes from and its abstraction into a comestible commodity is an important product of modernization."[61] While perhaps viewed by some as progress, it might be more specifically conceptualized as the progressive erosion of the relationship with food animals.

This progressive erosion is evidenced by changes in the presentation of meat for consumption. During the early modern period, not only was meat the focal point of the meal, the animal that was served at the table was recognizable; it was common for the entire animal to be served at the table, even with the head attached, and for the (male) head of the family to carve the meat.[62] Sociologist Norbert Elias notes that this practice changed during the Middle Ages; coinciding with an increased concern with civility, the presentation of entire animals for

Figure 24. Packaged bacon. (Source: istockphoto.com. Used with permission.)

consumption became less common. Elias argues that the civilizing process taking place at this time entailed refuting the animality of humans. One way people sought to distinguish themselves from their animal kin during this period was through "civilized" dining, which required obfuscating the animality of consuming other animals.[63] The distancing of humanity from animality was furthered during the Enlightenment period by a more general objectification of nature.[64] This separation process has been perpetuated not only by obscuring the animal origins of the individual animals we consume through methods such as trimming, deboning, and wrapping in plastic, but also by making the whole meat industry nearly invisible.

The industry is more than happy with this cultural arrangement. It understandably prefers to keep or even extend the distance between meat production/processing and the consumer. This desire is illustrated by the following statement by Peter Cheeke, a professor of animal science, in his textbook written for an animal science audience:

> For modern animal agriculture the less the consumer knows about what's happening before the animal hits the plate, the better. . . . One of the best things modern animal agriculture has going for it is that most people in the developed countries are several generations removed from the farm and haven't a clue how animals are raised and processed.[65]

Directly and indirectly, the industry has been able to further extend the distance between meat production, processing, and consumption through physical and symbolic distancing techniques.[66] The physical distancing takes place through attempts to make the process of turning animals into meat virtually invisible. As mentioned above, the meat consumers purchase contains few reminders of the animal it once was. For instance, the cellophane-wrapped bacon purchased at a grocery store (figure 24) does not resemble the species of animal it is derived from (figure 25). The vast majority of consumers do not see animals destined to become meat. The buildings that house the animals and those where animals are slaughtered are generally entirely nondescript, few people are allowed admittance, and recently industry-backed laws have been passed in some states to criminalize recording images from inside the industry because the distribution of undercover photos and videos threatens to bridge the gap between production, processing, and consumption.[67]

The techniques of symbolic distancing are plentiful. The chasm between production/processing and consumption is socially constructed and normalized by the industry, the government, and our culture. Gouveia and Juska remind their readers that "consumption and production . . . are analytically and historically linked processes underpinning social reproduction at its most fundamental level. The otherwise fictional separation between these two spheres is an artifact of power and socio-cultural as well as ideological construction."[68] Although it may be tempting to attribute this distancing solely to cultural developments, the industry and the political-economic system that are invested in current modes of production have also devoted significant energy to make sure that consumers do not confront the realities of the system. Consumers do not generally object to the blocking of these sympathies, whether it be by cultural, economic, or governmental forces, because it minimizes any discomfort involved.

Marketing has become a particularly important mechanism employed by the industry to create and sustain the symbolic distance between production, processing, and consumption.

Figure 25. Individual pig. (Source: istockphoto.com. Used with permission.)

Today consumers receive a great deal of information about the meat they consume through, and make decisions based upon, marketing. Industry marketing became increasingly important in the latter half of the twentieth century: as the health benefits of meat consumption came under increased scrutiny, the industry shifted from focusing on the material and instead began to focus on the symbolic, and accordingly increased their investment in marketing.[69] Up to this point, the industry had a secure consumer base, but now they needed to be more proactive because "once the connection between meat and robust health was questioned, there was very little to stop growing animal sentiments from further eroding the desirability of meat. However, meat eaters found that meat was socially acceptable in a more disguised, modest form and the meat trade was quick to exploit this."[70]

The new marketing strategy included marketing meat products based on taste, tradition, convenience, the "naturalness" of their production methods, and even on the basis of animal welfare.[71] It may seem paradoxical for the meat industry to address animal welfare in their marketing. However, with the health argument weakening and more people questioning the treatment of animals raised for meat, the industry has responded by constructing at least some of their products as animal-friendly. They have done so in several ways. One way is through the use of talking animals in their advertisements. The animals used are generally cartoon representations, such as Charley the Tuna. This type of advertising depicts the animals as having lived good lives and gives the impression that animals killed for their meat are willingly sacrificing themselves for the greater good of meat/dairy/egg consumption by humans.[72]

A related marketing strategy documented by researchers depicts happy animals living in idyllic country settings on family farms.[73] This strategy relies heavily on stylized photographs of livestock animals that give the impression that the small, family farms are alive and well. These animals are depicted as living happy, natural lives. These depictions couldn't be further from the reality of industrial animal production today. There is frequently a coterminous claim in these advertisements that the meat from these happy animals tastes better.

In addition to constructing positive messages, the industry does its best to spin the negative ones. When stories do emerge in the media about problems in the food production system, such as undercover footage of animals being abused or cases of food contamination outbreaks, the industry mollifies consumers by asserting that the individuals responsible are simply "bad apples." This not only gives consumers a sense of security, because presumably the "bad apples" are not the majority, it prevents them from questioning the system as a whole.[74] As I write this book, this strategy is currently being used to diffuse outrage over horse meat being combined with beef and consumed by unsuspecting people in Europe. Despite the massive scope of the problem, the industry is depicting it as an isolated problem. This case has not only illustrated how the industry here and abroad deflects critical attention, but also how upsetting the prospect is of consuming certain species of animals, a topic we turn to later in the chapter.

The distancing between production, processing, and consumption exploited and fostered by the industry is embedded in our culture, as evidenced in language. The language used by consumers, and promoted by the industry and even the government, obscures exactly what they are consuming and how it was produced.[75] In her excellent book exploring the role of language in the oppression of animals, Dunayer writes, "Deceptive language conceals the cruel conditions and treatment suffered by food-industry captives. Understatement, euphemism, positive description of negative realities, and outright lying hide the truth."[76] The language used to refer to the products is itself problematic. The terminology used to refer to meat from specific types of animals obscures the animal whose flesh is actually used. For instance, *beef* refers to the meat from cows, *pork* refers to the meat from pigs, *mutton* refers to the meat from sheep, and *chevron* refers to the meat from goats, instead of using more straightforward descriptions such as cow, pig, sheep, or goat meat.[77] Meat from birds is an interesting exception: the flesh from chickens, turkey, and ducks are generally referred to simply as chicken, turkey, and duck. There may not be a linguistic need for distancing consumers from these species, which are lower on the sociozoologic scale.[78]

Carol Adams introduced the concept of the *absent referent* to describe the process whereby the animal whose flesh is being used becomes virtually invisible, not only in a literal sense but also symbolically.[79] Through this process, people in western cultures can consume meat without confronting the animal it once was. It also works in connection with the *mass term*. Mass terms, such as *meat*, obscure the individuals involved. Failure to connect individual animals to meat means that one can consume *beef* and take solace in the fact that cows as a mass still exist. Individuation would be dangerous to the system and would make empathy more likely. Thus, livestock animals are not named: they are instead numbered, and once these numbered animals are taken to slaughter, a replacement animal assumes their number.[80]

This sanitized language is apparently not sufficient for the industry, as they have proposed further modifications. For instance, according to Dunayer, the National Cattleman's Association has urged industry members to replace the term *livestock industry* with the term *animal agriculture* because the word *industry* can have negative connotations. They find that the word *agriculture* is preferable because it evokes images of more traditional and historic farming, although the term *agriculture* has arguably lost some of its lustre. The industry also prefers the term *family farm* over *factory farm* for the same reasons. It also elects to use the term *euthanasia* to describe the killing of sick or surplus animals in production facilities, such as killing piglets by slamming their heads against objects. Croney refers to this use of "sterile language" by the industry as "misleading, ambiguous, or disingenuous."[81] It is, however, certainly good for business.

Sociologist Adrian Franklin refers to the distancing between production/processing and consumption discussed thus far as part of a network of avoidance rituals, which have been established as part of the modern use of animals as food.[82] The rituals promulgated by industry, the government, and the larger culture vis-à-vis the obfuscation of animal bodies in the purchasing process (e.g., the presentation of abstracted pieces of meat in grocery stores) and the consumption process (e.g., serving pieces of meat instead of the whole animal), as well as the removal of animal slaughter from the sight of the general public have already been discussed. What remains to be explored is how avoidance can manifest itself at the level of the individual consumer.

Philosopher Nancy Williams uses the term "affected ignorance" to describe individual-level avoidance or unwillingness among much of the public to understand where their meat comes from and whether or not their consumption is implicating them in an immoral process.[83] While a lack of knowledge, intentional or otherwise, may be a problem in some instances, it is important not to extrapolate from this that the masses are suffering from sheer ignorance. Many people are at least somewhat knowledgeable about the production of meat yet continue to consume it. They are not ignorant per se, but instead are able to neutralize this knowledge.

Knowledge can be neutralized in many ways. In attempting to understand the lack of reflexivity around industrial food production and consumption, Stuart and Worosz suggest looking to the techniques of neutralization originally explicated by Sykes and Matza in the 1950s to explain how individuals engage in deviant acts and are able to neutralize or distance themselves from their harmful actions. The techniques include denial of responsibility, denial of injury, denial of the victim, condemnation of the condemners, and appeals to higher loyalties. Although Stuart and Worosz are specifically interested in understanding these techniques vis-à-vis food safety issues, these techniques can also be usefully extended to understand the lack of reflexivity when it comes to the consumption of animals.[84]

Denial of responsibility is facilitated by the separation of production and processing from consumption in industrial animal agriculture. Consumers are so far removed from meat production and processing that responsibility for the death of the animal consumed is easily denied. *Denial of injury* is a little more difficult in this context, as adult consumers are aware that animals are indeed killed to create meat products. However, this rationalization technique is evident in the assumption made by many consumers that the existence of animal welfare laws keeps livestock animals free from injury and that their deaths are humane. The lives of animals in CAFOs are far from injury-free, yet industry marketing does give the impression that they are.

Denial of the victim is also difficult to perfect in this instance. This is one area where the linguistic techniques discussed earlier (e.g., use of terms such as *meat, beef,* and *pork*) are invoked. Joan Dunayer does an excellent job illustrating this process through her deconstruction of the linguistic practices surrounding the consumption of animals. She writes evocatively, "Flesh consumers deny nonhuman death. Avoiding direct reference to the bodies of murdered nonhumans, they say 'meat' rather than 'flesh,' 'muscle,' 'remains,' or 'corpse portion.'"[85] This denial of the victim is further perpetuated by the government, which refers to the animals in CAFOs as *units*, and the industry, which has communicated a preference for using the term *agriculture* over *livestock industry* because the former implies production based on needs instead of profits and makes no reference to the animals used within the industry.[86]

Condemnation of the condemners is another neutralization technique that can be used, although condemnation is certainly less common when it comes to meat consumption than the deviance that Sykes and Matza originally developed their list of techniques of neutralization to describe. Condemnation regarding meat consumption can, however, come from animal advocates and, increasingly, environmental and health advocates. In these instances, the condemners may themselves be condemned as impractical, overly emotional, or simply as do-gooders (my grandfather's personal favorite term). Condemning the condemners in this way makes it possible to neutralize the message by discrediting the messenger.

Finally, in *appealing to higher loyalties*, consumers of meat products can neutralize unease they experience by turning to justifications based on health and religion. Some people believe that it is unhealthy to eat a diet free of meat (although medical evidence is increasingly pointing in the opposite direction). Those readers who do not eat meat have probably heard something similar to this many times: "I'd like to be a vegetarian/vegan too, but I can't live without meat." If people genuinely believe that they physically cannot live without meat then they are able to at least partially neutralize condemnation of consumption: for them, at least theoretically, it is necessary for survival. Religion is another higher loyalty that can be appealed to. This argument generally runs along the lines of "I eat meat because that's what God put animals on earth for." Of course, this logic depends upon which "God" is being referred to and is vulnerable to many critiques. It would take us too far afield from the purpose of this book to discuss those critiques here; however, those interested in these specific critiques would likely find the works by Scully, Kalechofsky, Foltz, Phelps, and Rosen useful.[87]

Although these neutralization techniques employed by individuals, along with the avoidance rituals promoted by the industry, government, and embedded culturally, are impressively effective, there can be fractures that allow acknowledgement and discomfort to percolate to the surface. For instance, cases of meat contamination and subsequent recalls can begin to lift the thin veil covering the production of meat.[88] Or citizens might encounter a news report in the media detailing abuses uncovered inside of a CAFO. If one is not able to use the rationalization techniques detailed above to neutralize this new information, s/he may experience unease, which the psychological community commonly refers to as *cognitive dissonance*.

Psychologists use the term *cognitive dissonance* to refer to the psychological discomfort that emerges when one cannot fully neutralize an inconsistency in their cognitions. In developing the concept, Leon Festinger posited that people in a state of cognitive dissonance try to reduce the dissonance (because it causes discomfort) and will avoid contexts and information that contribute to the dissonance.[89] This avoidance of contrary information is evident in decision making that impacts animal well-being. For instance, researchers conducted a study on the consumption of animal products in Great Britain, Italy, and Sweden and found that *only half* of respondents who indicated they believed that animal welfare is *very important* consistently considered animal welfare when making purchasing decisions. The authors conclude, "consumers who care about animal welfare suffer cognitive dissonance from livestock product consumption and so may prefer not to think about welfare when they are buying meat."[90]

Attitudes and behaviors related to consuming animals are therefore complex and are impacted by the larger culture, stakeholders, social institutions, and psychological strategies. They are also affected by experiential and social structural variables. For instance, on average race, class, gender, and interactions with farm animals affect attitudes toward animal well-being. Research has documented that where a child was raised has the greatest impact on attitudes about the well-being of animals, such that those who grew up on farms report being least concerned with animal well-being. Additionally, women, the economically disadvantaged, less educated, younger and middle-aged people, and Blacks exhibit higher levels of concern for animals. The authors argue that their findings support the "underdog" hypothesis that those who are socially disadvantaged empathize more with the plight of animals.[91]

Kalof and colleagues find in their examination of a random sample of U.S. adults that the impacts of structural/demographic factors on vegetarianism may be mediated by values and beliefs. They report, for instance, that the effect of gender loses significance once values and beliefs are controlled for. This suggests that the gendered difference in vegetarianism may be related to women holding different values and beliefs than men. Specifically, they find that altruism has a significant positive effect on vegetarianism. Their findings are also consistent with the underdog hypothesis; specifically, they find that the moral views of subordinated populations (i.e., women, racial/ethnic minorities) are similar, and the views of white men are what stand out as being unique.[92]

The findings of these two studies stand in contrast to the postmaterialist theory that proposes that those individuals who are concerned about issues that do not affect their immediate well-being, such as environmental and animal issues, have the luxury of being able to be concerned about these issues because they have enough resources that they do not have to worry about their immediate well-being. The social movements associated with environmental and animal issues may in fact be more heavily populated by those with resources, but concern about these issues appears to be stronger among those groups of people that have experienced subordination.[93]

There are therefore many factors that affect consumptive behavior, which has by extension contributed to a wide array of practices vis-à-vis meat consumption and perspectives on what the place of animal-derived products in the human diet should be. However, the diversity of values, attitudes, beliefs, and behaviors in the United States narrows considerably when the specter of consuming species of animals who are situated somewhere between the categories of livestock and pet, such as horses, arises. This tells us as much about our relationship with animals we do consume as it does about our relationship with animals we will not consume.

WHEN AND WHY SPECIES MATTERS

The increasing need for avoidance rituals and neutralization techniques around meat production/processing/consumption is likely being fueled at least partially by a growing tension within our culture. According to historian Richard Bulliet, there has been an important shift in our relationship with animals. He asserts that we are in the midst of the *postdomestic era*, which began in the 1970s and is characterized by people in developed countries becoming physically and psychologically more removed from the animals they consume. Simultaneously, people have become closer with their pet or companion animals. This era is contrasted with the previous, *domestic era*, within which people had daily contact with animals, including those that would later be consumed. Such contact was much less commonly affective. According to Bulliet, while meat consumption was normalized in the *domestic era*, in the subsequent *postdomestic era* a tension emerges between the strong, positive feelings people have for pet species and their consumption of other animals. He explains,

> a postdomestic society emerging from domestic antecedents continues to consume animal products in abundance, but psychologically, its members experience feelings of guilt, shame, and disgust when they think (as seldom as possible) about the industrial processes by which domestic animals are rendered into products and about how those products come to market.[94]

This tension is no doubt exacerbated by the fact that the number of animals being produced for human consumption is growing and the quality of their lives is declining at exactly the same time that the love of pets is increasing. We might expect this tension to worsen and efforts to neutralize it to increase over time as the number of animals fed through the system continues to increase and affection for pet animals grows. This tension may escalate in the largest meat producing countries in particular, including the United States, Canada, Australia, and New Zealand, which ironically also have the "strongest postdomestic mentality."[95]

Bulliet's identification of a *postdomestic era* characterized by a growing tension between the treatment of animals deemed pets and those deemed consumable raises a fundamental question: Why do people love some animals and consume others who share many of the same characteristics? Focusing on exactly this question, Hal Herzog titled his recent book, *Some We Love, Some We Hate, Some We Eat: Why It's So Hard to Think Straight about Animals.*[96] In beginning to address this question it is worth noting that there is not an absolute divide between the species of animals that people will and will not consume. The line between consumable and taboo food choices varies across time and space, making it appear somewhat arbitrary.[97] Nonetheless, cultures can reach near consensus about many species, as evidenced by the case of the horse.

In 2013, the news that horse meat had been sold as beef to consumers in Europe seriously upset many people. Yet archeological evidence indicates that historically horses were consumed in northern Europe. This consumption, however, dropped off in the later Middle Ages and subsequently became taboo. There is no agreed upon definitive reason why consumption of horses, or of other animals for that matter, became uncommon or even taboo. The proposed reasons include that during the Middle Ages people came to rely on horses for their labor power, which may have made them more valuable for this purpose than as food; as working animals, they came to be revered in western countries, even becoming symbols of nobility, and escaped industrialized production and consumption.[98]

Yet horses have not entirely escaped the slaughterhouse. Although federal regulation of horse slaughter ceased in 2007, effectively ending horse slaughter in the United States, since that time horses have simply been sent from the United States to Canada for slaughter. Horse slaughter in Canada increased by 50 percent after the practice was ended in the United States. In 2008 alone, Canada slaughtered over 110,000 horses and produced approximately 14,000 tons of horsemeat, making it one of the largest producers globally.[99] Horses may have thus far escaped industrial production, but not industrial slaughtering. And thinking about horses being sent for slaughter, illustrated in figure 26, to date elicits much more discomfort than the sight of chickens, for instance, being sent to slaughter, pictured in figure 27.

Consumption of dogs also has a history of being taboo in western cultures. A number of interrelated reasons may account for this specific taboo. First of all, dogs may not have been considered a suitable food choice because they were considered dirty and impure. Refusal to eat animals considered dirty or impure may have served to symbolically distance humans from animality; like the shift away from presenting the full animal's body at the table, not consuming "impure" animals could frame humans as more careful consumers than instinct-driven

Figure 26. Transport of horses originating in Tennessee to slaughter. (Source: J. Caramante, Equine Welfare Alliance/Wild Horse Freedom Federation Investigator. Used with permission.)

Figure 27. Transport of chickens to slaughter. (Source: People for the Ethical Treatment of Animals. Used with permission.)

animals. Second, and relatedly, dietary prohibitions on consuming certain species, including dogs, may have been a mechanism for religions to "civilize" and discipline people. Third, there is some speculation that dog consumption became taboo because dogs had been involved in pagan rituals and took on an antireligious character. Fourth, there may have been hygienic concerns about consuming dogs because they carried bacteria and infections. Fifth, the food that dogs themselves consumed may have rendered them unsuitable for human consumption. They are capable of consuming blood, the remains of dead animals, and even human remains, and therefore dogs, along with carnivores more generally, might have been considered unacceptable as food. Finally, dogs, like horses and cats, are considered liminal animals—they come (uncomfortably) close to the human side of the perceived animal/human divide.

Actual and symbolic closeness of these animals with people may have rendered them unsuitable as food. There is a universal taboo against harming members of one's social group. Therefore, people who are socialized into a culture that considers certain species as family members may be disinclined to consume them.[100] Although there is a tendency to emphasize this possibility that people avoid harming animals they consider kin, it is also the case that people avoid becoming close with animals who are going to be harmed.[101] Regardless of the reason for their existence, these food taboos have protected some species of animals from arguably the worst human system—the food system—for animals to find themselves in (although other systems are harmful for animals as well, such as the medical, entertainment, and shelter systems).

The irony regarding our variable treatment of different species is that the species we treat poorly (i.e., livestock animals) are the ones that actually benefit us the most economically and occasionally as sources of labor power, while the animals we treat well and even spoil (i.e., pet animals) are of limited instrumental value.[102] This may seem odd to an observer; however, it is the economic value of livestock animals that has put them at greatest risk of treatment designed to maximize profits. The fact that the services that pet animals provide are primarily noneconomic (e.g., companionship, affection) has protected them from the worst and most widespread (ab)uses, but certainly not from all of them. While the economic value of livestock animals can serve to distance the general public from them, the noneconomic/affective value of companion animal species has drawn us closer to them and, potentially more transformatively, toward animals more generally.

Literal and symbolic distance is the key to making another being consumable. Although as discussed earlier, there are words used to distance ourselves from the animals we consume (e.g., meat derived from pigs is referred to as *pork*), there are no similar linguistic distancing mechanisms in the English language for horse, cat, and dog meat.[103] Physical and symbolic proximity to these animals has contributed to food prohibitions, which have been reinforced by the attachment of a significant social stigma to their consumption. Consuming taboo animals can be denigrated as a sign of poverty or even treated as deviant and illegal.[104] For instance, California enacted a law prohibiting the consumption of animals "commonly kept as a pet or companion" in response to a case where two Cambodian immigrants killed and consumed a dog.[105] As illustrated by this case, attention in the United States is willingly focused on the inexcusability of consuming certain species of animals while unquestionably accepting the consumption of others.

The potential tension caused by the discrepant treatment of animals (i.e., treating some as family and eating others) has thus far been mitigated by the erection of an arbitrary barrier between species of animals and refusing to reflect upon the nature of this barrier.[106] To appreciate the power and scope of this barrier, one merely has to "imagine the level of public outrage if it were disclosed that the government was subsidizing the factory farming of puppies and kittens."[107] This discrepancy is so profound that it even impacts the perceptions of the professionals who are charged with caring for animals. Studies of veterinary students have found that the students' assessments of the acceptability of performing painful procedures on animals varies depending upon the species of the animal in question; they are more likely to find it acceptable if the species in question is a livestock animal.[108]

James Serpell explains that the solution to the discrepancy between our attitudes toward and treatment of food and pet animals employed by those who work with animals and the general public alike has been "to compartmentalize—to allocate our moral obligations to some animals but not to others—and to invent elaborate belief systems and 'just-so stories' to explain why animals do not actually matter even when our gut instincts, our moral intuitions, tell us that they do."[109] This cultural compartmentalization has significant consequences for those animals we are taught not to care about, but this harm is not necessarily limited to animals. As Serpell goes on to explain, these cultural techniques have a long history of harming humans as well: "These are, of course, precisely the same techniques that people have used throughout history to justify the abuse and persecution of other humans, and that, more than anything else, is why animals are a social issue."[110]

Another way of attempting to resolve the tension caused by our discrepant treatment of animals along species lines while simultaneously justifying their subjugation is to trivialize or ridicule pet keeping. This technique serves to distance humans from animals entirely: it is not necessary to explain why some animals are perceived differently than others if they can all be marginalized equally.[111] On the whole, however, this technique does not appear to be working, as the United States is increasingly a nation of pet lovers. Serpell suggests it is this growing cultural affection for pet animals that could actually be the undoing of industrial animal agriculture. He articulates this possibility as follows:

> when we elevate companion animals to the status of persons; when we empathize with them and acknowledge their resemblance to ourselves, it becomes obvious that the notion of human moral superiority is a phantom: a dangerous, egotistical myth that currently threatens our survival. Ironically, as the forerunners of animal domestication, pet-keeping led us into our present, destructive phase in history. Perhaps, by making us more aware of our biological affinities with animals and the natural world, it will help to lead us out again.[112]

Increasing levels of pet ownership and affection for pets, coupled with increasing numbers of animals being intensively raised for meat, may exacerbate the tension Serpell describes above, which Bulliet refers to as the postdomestic tension. I am dubious about the possibility that this tension will be enough to dramatically change meat consumption patterns; as discussed in this chapter, a number of mechanisms have been developed by industry interests, the government, culturally, and at the level of the individual to mitigate this tension. This tension, coupled with increasingly pressing negative impacts of meat production, processing, and consumption, however, might just be sufficient to induce change.

Industrialization Fallout

As discussed in the previous chapters, the industrialization of the production and processing of meat made it widely available for consumption. As a result, today meat is plentiful and relatively inexpensive to purchase, and there are many varieties available. The USDA has touted these developments, along with the corresponding efficiencies in production and processing, as advantages of the industrialization of meat production and processing.[1] This and other celebrations of industrial animal agriculture, however, take place without consideration of the externalities created by the industry and the subsidies it directly and indirectly receives that make it appear so productive and efficient.[2]

The purpose of this chapter is to explore the most consequential of the externalized effects of the industry to provide the much-needed larger context for evaluating the contribution of the industry. In other words, this chapter adds the impacts of the industry to the chain linking production with processing with consumption (pieced together in the previous chapters), and in so doing brings the respective links in the chain closer together. I am certainly not the first to assert that there is a need to conceptually and logistically tether meat production, processing, and consumption with their associated consequences. For instance, Carol Adams argues we "must address the fact that our meat-advocating culture has successfully separated the *consequences* of eating animals from the *experience* of eating animals."[3] What this chapter does is contribute to this important goal by examining the wide array of industry impacts and consequences of the consumption of its products, including the negative effects on animal welfare, worker health and safety, human health, the environment, food security, the communities where production and processing facilities are located, and even the larger culture. I conceptualize these impacts as varying along a continuum from direct to indirect, and the chapter traces this continuum, beginning with the most direct impacts, which are felt by the animals whose bodies are produced and processed by the industry.

ANIMAL WELFARE

As detailed in previous chapters, the economic and technological changes that have shaped animal agriculture have also dramatically altered the lives of animals in the system. In the traditional husbandry of the past, producers had vested interests in the well-being of the animals because harm experienced by the animals also harmed the producer in the form of diminished returns; technological fixes to mitigate these diminished returns were not yet commonplace. Rollin describes this traditional relationship between animals and producers as a mutually beneficial contract.[4] It is important to keep in mind, however, that even though harms endured by livestock animals might have been more closely tethered to the well-being of producers than is currently the case, there still was an inherent power imbalance (a more benevolent power imbalance perhaps, but a power imbalance nonetheless).

There is a relationship between increasing industrialization in animal agriculture and an exacerbation of this power imbalance, which is accompanied by greater animal welfare challenges. This relationship has not been perfectly linear: interventions by the growing field of animal welfare science and pressure exerted by consumers and advocacy groups have periodically contributed to improvements in animal welfare. These interventions, however, are always within the strict confines of maximizing profits and have therefore been limited because, as Matheny and Leahy observe, "when animal welfare competes with economics, economics usually wins."[5] The modest welfare improvements that have been made along the way have been outpaced by mounting animal welfare concerns ushered in by increasing industrialization and consolidation.[6] As a result, "our domesticated livestock have never been as cruelly confined or slaughtered in such massive quantities in all of history."[7]

Space constraints do not permit an examination of all current welfare challenges in the industry. Instead, we will focus on the three main areas of concern usefully delineated by Rollin.[8] First of all, current production methods are causing what are referred to as *production illnesses or diseases*. Due to specialized breeding, feeding, growth promoters, and antibiotics, animals produced for food grow much faster and larger than their ancestors did. This rapid and extreme growth places excessive strain on them, particularly their cardiovascular systems and skeletons.[9] For instance, one study found that 90 percent of chickens produced for meat today suffer from gait abnormalities.[10] Further, selective breeding of pigs for traits that increase production and value has contributed to numerous characteristics and conditions that negatively impact their welfare, including increased anxiety, aggression, leg problems, heart disease, and porcine stress syndrome, which like "flip-over syndrome" in chickens, causes them to die suddenly. In short, these animals are bred for profitability, not for longevity and welfare.[11]

These physical problems, coupled with the environment inside CAFOs, contribute to weakened immune systems among livestock animals. Combined with the close quarters these animals live in, it is a perfect storm for illness and contagion. For instance, industrially produced cattle are vulnerable to respiratory diseases and digestive problems. Respiratory and airborne diseases also commonly afflict pigs in confinement, whereas poultry are most vulnerable to parasites, infections, and metabolic diseases.[12] Overall there are relatively high preslaughter death rates in CAFOs, ranging from 4 percent for cattle, to 14 percent for pigs, and 28 percent for some chicken varieties.[13] These losses are factored into the costs of production. The suffering these animals endure as they die is a welfare concern, as is the euthanasia of ill animals in CAFOs and slaughterhouses. The most common forms of euthanasia used in the industry include carbon dioxide poisoning, blunt force trauma, and cervical dislocation. The humaneness of these methods is debated.[14] Given the environment these animals live in and the high mortality rates, their plight has been described as remaining "in a state of dying until they're slaughtered," one way or another.[15]

The problems detailed above are compounded by the second negative animal welfare impact of industrialized production Rollin identifies: in industrialized production, *individual husbandry no longer takes place*. There are simply too many animals and too few workers to attend to the well-being of individual animals. Therefore, it is impossible to identify individual animals suffering from the illnesses discussed above or experiencing stress from things like air quality problems caused by ammonia, heat, physical mutilations (e.g., debeaking, castration, tail-docking), and in some cases, withdrawal of food and water to increase production.[16] The inability to provide individual husbandry and redress the stress exhibited by individual animals no doubt contributes to the high mortality rates.

The final welfare concern Rollin explicates is *physical and psychological deprivation*. In the extreme confinement of industrial animal agriculture, many animals experience total confinement. The most significant restrictions are experienced by laying hens kept in battery cages, sows kept in gestation crates, and dairy cattle kept in stalls. This degree of confinement and behavioral restriction causes mental distress, abnormal behavior (e.g., cannibalism), and physical problems such as sores and muscle wasting.[17] Cattle produced for meat are also affected by conditions on feedlots, such as standing in their own manure, being exposed to weather extremes, and human handling. In general though, cattle have not endured the extreme deprivation that other livestock animals have. Mench and colleagues suggest that this may change in the near future, as they speculate that feedlots will soon relocate to the Midwestern or the eastern states to capitalize on the low cost of feed derived from ethanol production there, and as part of this transition cattle producers will adopt more intensive practices, including keeping them indoors.

Although Rollin focuses on welfare issues arising from industrialized production, it should be acknowledged that the industrialization of processing has also given rise to animal welfare concerns. For instance, because of the increasing speeds of disassembly lines in slaughterhouses, stunners have to work very quickly, which compromises precision and results in some animals being processed/dismembered while conscious.[18] There are also numerous welfare issues that arise in the transport of animals from production to processing, although due to space constraints, that is terrain that cannot be covered here.

Given all of these potential animal welfare issues, one may reasonably wonder what checks are in place to monitor animal welfare and provide basic protections for farm animals. The most succinct and honest response to such a query is "it's complicated." How to best measure animal welfare and what the goals of animal welfare should be are both contested issues, and practices vary across countries. For instance, in the United Kingdom, a list of five freedoms that must be present to ensure animal welfare was drafted by a government-appointed committee of experts. These five freedoms include freedom from hunger and thirst; freedom from discomfort; freedom from pain, injury, and disease; freedom from fear and distress; and freedom to engage in normal behavior. No such declaration exists in the United States.

Measuring welfare attributes also varies. Animal welfare in the European Union is measured relatively holistically, and includes health, productivity, physiology, and ethology indicators. The United States, conversely, officially uses only health indicators.[19] This means that in the United States, even though individual animal scientists may employ numerous measures of welfare in their research, poor animal welfare in the industry is technically viewed as manifesting itself only in negative health outcomes.

Much like the measures of animal welfare, the laws governing animal welfare in the United States are limited. There are no federal laws regulating the treatment of animals raised in animal agriculture. The Animal Welfare Act, which many think would protect livestock animals, explicitly excludes farm animals from its purview. State anticruelty laws are also of little help. Most state-level anticruelty laws exclude actions against farm animals that are accepted as customary, and some further exclude farm animals from protection. The federal-level Humane

Methods of Slaughter Act does regulate slaughter; however, it excludes poultry and fish (which comprise 98 percent of slaughtered animals) as well as ritual slaughter, and what the law does cover is weakly enforced.[20] The general conclusions that can be drawn about these laws are that they exclude those most in need of protection and "as long as it is sufficiently common, virtually any farming practice, no matter how painful to animals, may be legal in most of the United States."[21] Thus, the well-being of livestock animals is severely compromised within the current legal, political, and economic regimes.

WORKER SAFETY AND WELL-BEING

The impact of industrial animal agriculture on worker safety and well-being is fairly localized. These effects are felt specifically by workers employed in both the production and processing sectors of the industry (although the effects can certainly also indirectly impact those with close relationships to the workers as well). In general, workers in the industry have been experiencing stagnating wages, high illness and injury rates relative to other industries, and other problematic impacts tied to the unique nature of their work.

There are currently approximately 79,370 slaughterhouse workers and 29,570 animal agriculture workers employed in the United States.[22] Between 1980 and 2000, while profits in the industry more than tripled, the wages of workers stagnated.[23] As of May 2012, the median hourly wage of animal agriculture farm workers was $10.61 (mean annual wage of $24,040) and for slaughterhouse workers was slightly more at $11.70 (mean annual wage of $24,930).[24] These wages are barely above the poverty line, set at $23,050 per year for a family of four in 2012 by the U.S. Department of Health and Human Services.[25]

In order to further contextualize this level of monetary compensation it is important to also take the nature of the work into consideration. The industry has relatively high injury and illness rates. Estimates from the year 2011 put the incident rate of nonfatal injuries per 100 full time employees at 6.4 for animal production and 4.8 for animal slaughtering. These rates are higher than those in the notoriously dangerous mining and construction industries, which were 2.9 and 3.8 respectively in the same year.[26]

Even though the official injury and illness rates in the industry are high, they are likely conservative estimates due to underreporting. As mentioned earlier, the industry employs a high proportion of immigrant laborers and undocumented workers. The U.S. Immigration and Naturalization Service even indicted Tyson Foods for smuggling illegal workers in to work in slaughterhouses.[27] These workers are particularly vulnerable and are understandably less likely to report injuries and illnesses for fear of losing their jobs.[28]

Regardless of the exact number of injuries and illnesses that occur in the industry each year, it is apparent that the workers confront a number of significant risks. More specifically, workers in the animal production segment of the industry are at the front line of exposure to airborne pathogens and bacteria. More than one hundred studies link the air inside CAFOs to lung disease, nausea, depression, brain damage, and even death among workers.[29] It is further estimated that at least 25 percent of CAFO workers are afflicted by respiratory diseases.[30]

Injuries are prevalent in the animal slaughtering segment of the industry, where there is a perfect storm of contributing factors, including high disassembly line speeds, workers using sharp instruments, improper stunning of some animals, and inadequate worker train-ing.[31] Currently, cutters on the disassembly line have approximately three seconds to cut each piece of meat.[32] Such quick, repetitive movements not only contribute to repetitive movement disorders, such as carpal tunnel syndrome, they also make it more difficult to prevent injuries inflicted by the sharp tools or by improperly stunned, thrashing animals.[33] Resultant injuries are described in Occupational Safety and Health Administration accident reports as follows:

> Employee hospitalized for neck laceration from flying blade. Employee's eye injured when struck by hanging hook. Employee's arm amputated when caught in meat tenderizer. Employee decapitated by chain of hide puller machine. Employee killed when head crushed in hide fleshing machine. Caught and killed by gut-cooker machine.[34]

Even in the absence of such extreme incidents, the work environment can be challenging. For instance, there have been reports of workers routinely being denied bathroom breaks, and Cook recounts the story of a pregnant woman with morning sickness who was instructed not to leave the line and instead told to vomit in a garbage can.[35]

The combination of the work environment and stagnating compensation have resulted in high employee turnover rates in the industry, as high as 200 percent per year in some facili-ties.[36] Observing these high turnover rates, Cook aptly remarks that the workers "don't last much longer than the pigs."[37] This high turnover in the industry makes it difficult to retain well-trained people, which is necessary if safety and animal welfare are going to be improved. This problematic cycle needs to be broken.[38]

The work environment inside slaughterhouses, which workers and their unions in the Netherlands have referred to as "brutalizing," is beginning to receive critical attention by the media.[39] In 2005, the *New York Times* published an editorial critiquing the industry and demanding that attention be paid to the plight of the workers, which read in part, "what is most alarming at the slaughterhouse is not what happens to the animals—they have already met their fate. It is what happens to the humans who work there."[40]

Increasingly, academics are also weighing in and drawing attention to the impacts on workers beyond just physical well-being. In their work, Eisnitz, Fink, and Rémy point to a serious contradiction inherent in the industry that can have emotional and psychological con-sequences for workers.[41] On the one hand, due to humane slaughter laws workers are expected to ensure a humane slaughter. On the other hand, they are required to follow the ethos of the workplace and conduct their work as quickly, not as humanely, as possible. Not surprisingly, this contradiction—recognizing animals as sentient beings while processing them as a form of property—can be difficult for some people to reconcile.

The experiences of academics who have worked inside slaughterhouses provide evidence of the emotional and psychological toll this work can take. For instance, anthropologist Deborah Fink worked inside a slaughterhouse and describes the effect the work had on her as follows:

> Before I left the [slaughterhouse] plant I was carrying heavy depression and thoughts of suicide. . . . I found that my mind would fill with grotesque flashbacks, and I was unable to process events or emotions as I had before. I dreamed about looking into a combination of meat [containers cut up meat is thrown into] and seeing detached arms and tormented faces reaching up to me to be saved or to pull me in.[42]

These negative emotions may be turned inward or outward.

There is anecdotal evidence of animal abuse by some slaughterhouse workers.[43] In rare instances such abuse receives media attention. In one such instance, two workers at a slaughterhouse in California were captured on undercover video in 2008 kicking, shocking, and dragging "downed' (i.e., injured or disabled) cattle. The president of the company that owned the facility said in a statement that he was "shocked and horrified" by the actions of these two individuals.[44] The president's statement implies that these are a couple of "bad apples" and that this is an individual and not a systemic problem; however, it is not self-evident that is the case.

In addition to such anecdotes, research is beginning to document the potential emotional and psychological consequences of working within the industry. Although the research in this area is still limited, it points to some potential disturbances caused by working in slaughterhouses in particular. In one of the earlier pieces of research, Herzog and McGee studied students on a college work crew responsible for slaughtering cattle and pigs to assess their reactions to their work. Forty-eight percent of their sample reported having dreams about killing animals, and 19 percent reported sleep disturbances after their first killing experience.[45]

Another study conducted in Turkey compared the psychological well-being of forty-three butchers working in slaughterhouses and thirty-nine butchers working in supermarkets with a control group of office workers. The researchers report that the butchers have statistically greater experiences of somatization, obsessive-compulsiveness, interpersonal sensitivity, depression, anxiety, anger-hostility, phobic anxiety, paranoid ideation, and psychotism than the control group. Importantly, however, it is unclear if these problems developed for the butchers in the course of their work or if these problems predated their employment as butchers.[46]

MacNair's work on perpetration-induced stress among soldiers, executioners, and law enforcement officers, detailed in her book *Perpetration-Induced Traumatic Stress: The Psychological Consequences of Killing*, can help to theoretically contextualize the findings of these studies and the anecdotal evidence. In her book, MacNair argues that the perpetration of violence, even that which is socially approved of, can have significant negative psychological consequences for the perpetrators, even when they were just following orders in perpetrating the violence. Although MacNair has not had an opportunity to study slaughterhouse workers, she suggests that they be examined for perpetration-induced traumatic stress. She poses the following questions worth considering: "Does the fact that these are merely animals prevent the psychological consequences that would accrue if people were to be treated in this way? Does the fact that this kind of violence is done in massive numbers make it more of a psychological problem than violence to one or a few animals would?"[47] Research is needed to address these questions. If the industry does have negative psychological impacts on workers, it might manifest as cruelty and be directed at animals within the industry, or, as some have theorized, it could spill over into the broader community—a possibility discussed later in the chapter.

HEALTH IMPACTS AMONG THE GENERAL PUBLIC AND CONSUMERS

While certainly not everyone is concerned about animal welfare issues or even the well-being of the workers inside the industry, human health issues arising from current processes of meat production, processing, and consumption can potentially capture a much wider audience. One way to begin to appreciate these impacts is to look at how the major illnesses afflicting people have changed over the past century against the backdrop of changing production, processing, and consumption practices. At the turn of the twentieth century, approximately 40 percent of deaths were attributed to infectious diseases (e.g., typhoid) and only 16 percent were caused by what are referred to as diseases of affluence (such as cancer, stroke, and heart disease). By the early 1970s, the tables had turned and infectious diseases were causing only 6 percent of deaths, whereas the diseases of affluence were now responsible for 58 percent.[48] Certainly, many things changed over time that could have contributed to this shift, such as changes in hygiene and medical technologies. Yet there is significant evidence that the increase in the diseases of affluence is linked to the coinciding dramatic increase in meat consumption during this time.[49]

It is not just the increase in meat consumption that has been linked to increasing rates of diseases of affluence; changes in the way meat is produced in the second half of the twentieth century have also been implicated. Some substances used in the production of meat have been identified as potentially harmful. For instance, six growth hormones used in cattle have been linked to breast, prostate, and colon cancer in human consumers. In response to these findings, the European Union banned the use of growth hormones in cattle, but the U.S. government has not. Another problem related to production methods that has been identified is that CAFO-produced beef has higher levels of fat than pasture-raised, grass-fed. Saturated fat raises the problematic type of cholesterol and can lead to heart disease. It is also linked to diabetes and cancer. The main sources of saturated fats in the U.S. diet are animal products.[50] Rolling consumption levels back to earlier rates will not therefore in and of itself resolve the associated health problems because the way that meat is produced has changed so dramatically. The meat is different, and so are the risks. The evolving risks, particularly in the form of food-borne illnesses, prion diseases, influenza, and antibiotic resistance, are the focus of this section.

Food-Borne Illnesses

Food-borne illnesses are estimated to cost the United States between $30 and $60 billion in medical costs per year.[51] Bacterial contamination has been a particularly persistent problem in the meat industry. The irony, as Mallon articulates it, is that given all of the scientific and technological tools that have been brought to bear upon meat production and processing, detailed in previous chapters, "the U.S. meat supply should be among the safest and most sanitary in the world. Unfortunately, this is not the case today and unsuspecting Americans and other consumers are ingesting meat that can more properly be labeled a biohazard."[52] To say that it constitutes a biohazard is perhaps overstating it, but Mallon's statement does raise an important question: Given the safety procedures and technological fixes that have been applied to producing the meat supply, why do these problems continue to exist? The quick answer is that the structural problems have not been addressed. Before discussing how these underlying issues may be redressed, however, a survey of the manifestations is necessary.

The Centers for Disease Control and Prevention estimates that each year approximately 48 million people in the United States (or 17 percent of the population) are sickened by a food-borne disease, 128,000 are hospitalized, and 3,000 die.[53] The top five pathogens that cause food-borne illnesses resulting in hospitalization are, in descending order, *Salmonella* (35 percent), *Norovirus* (26 percent), *Campylobacter* (15 percent), *Toxoplasma gondii* (8 percent), and

Escherichia coli 0157:H7 (4 percent).[54] Four of the five foods most commonly identified as being responsible for food-borne illnesses are animal-derived products, specifically, eggs, poultry, seafood, and beef.[55] Not only is salmonella the most common cause of food-borne illness resulting in hospitalization, these infections are on the rise, and as of 2010, the incidence was three times greater than the "national health objective target."[56]

Although E. coli is not the most common pathogen, it has been well studied, and these studies provide insight into how pathogens can enter the meat supply. E. coli is contained in the intestines of cattle and sheep and is more common in those fed grain, which has become the standard in industrialized production. E. coli does not make the animals who house the bacteria sick, but it can sicken people who come into contact with it. Between 5 and 10 percent of people infected will develop hemolytic uremic syndrome, which can lead to kidney failure, seizures, paralysis, and death.[57]

E. coli enters meat when manure and/or the contents of an animal's gut come into contact with the carcass during the slaughtering process. Keeping them out of contact is made difficult for slaughterhouse workers due to the increasingly fast disassembly speeds. It is also complicated by the fact that there is an exceptionally high turnover rate in the industry, which means that workers are less likely to develop the skills necessary to keep manure and gut contents away from carcasses.[58] There has also been a relaxing of standards around manure. In the past, livestock that had been contaminated with manure was not added to the food supply. Now it is considered a "cosmetic blemish," and it is rinsed off and included in the food supply.[59]

Consumption of contaminated meat puts consumers at risk of being sickened by E. coli, but direct consumption is not the only possible means of contracting E. coli. It can also travel from animal manure into the water and contaminate crops. For instance, in 2006 there was a huge spinach recall because it had been contaminated by E. coli, which had originated at a CAFO and had spread to the spinach via runoff.[60]

As discussed earlier, new regulations requiring bacteriological testing were implemented despite legal attempts by the industry to block the new requirements. The new regulations, however, have had a limited effect on controlling food-borne illnesses for two main reasons. First, instead of addressing the root problem, the industry has implemented technological fixes in the form of decontamination technologies such as rinsing, ozonation, steam pasteurization and vacuuming, and irradiation that have been unable to entirely eliminate contamination.[61] The larger structural issues that facilitate contamination through amplification have not been addressed. Juska and colleagues explain that while bacteriological testing has reduced the amount of bacteriological contamination at the front end of the processing, and no doubt the decontamination techniques help at the back end, industrialization and concentration in the industry amplify the contamination that does survive these processes. Today meats are distributed more widely than in the past; many more animals go through the slaughterhouses, potentially spreading contamination; and meats from more animals are combined together. In fact, the USDA estimates that one infected animal can contaminate sixteen tons of ground beef.[62]

Second, these regulations are generally enforced in a limited way or through self-regulation.[63] Joy vividly summarizes the consequences of industry self-regulation as follows: "What this all comes down to is that corporations, whose primary objective is to increase their profit margins, are left to police themselves. We have left the fox to guard the chicken coop. And not surprisingly, we have ended up with shit in our meat."[64] Fecal matter, however, is not the only risk to human health that has emerged from the system.

Prion Diseases

Industrialized animal agriculture has also facilitated the transmission of prion diseases, or transmissible spongiform encephalopathies (TSEs). These are neurodegenerative disorders, and they can be transmitted across species lines.[65] The most well-known prion disease is undoubtedly Creutzfeldt-Jakob disease (CJD) in its human variant and bovine spongiform encephalopathy (BSE), or mad cow disease, in its animal variant. This illness came into public consciousness in the 1990s when it was revealed that people who consumed BSE-infected cattle were contracting fatal CJD. Since the initial outbreak, which was concentrated in the United Kingdom, there have been approximately 190,000 cases of BSE reported.[66]

Although the focus has been on the transmission of the disease from cattle to humans, there is speculation that the illness actually originated in humans and was transmitted to animals beginning in the 1950s when human remains from Hindu funerals infected with CJD intermingled with animal bones and were ground up and sent to England and used in livestock feed.[67] For years, the industry had been adding animal remains to livestock feed as an inexpensive way to increase the protein content with the goal of making livestock grow larger, faster. The animal remains rendered and added to livestock feed reportedly included those that had been considered unfit for human consumption along with road kill, dead horses, and euthanized cats and dogs. This practice gained attention once it was determined that BSE was being spread among cattle by the practice of feeding cattle by-products to other cattle. Through this practice, cattle were being changed not only from herbivores to carnivores, but even to cannibals.[68] The mad cow crisis frightened and angered the public and in so doing brought the connections between production and consumption into closer focus. The saying that "you are what you eat" suddenly gained new meaning, but a new lesson also emerged: it turns out, "we are what what we eat eats too."[69]

Citizens in England and other countries, including the United States, turned to their governments to protect them from these newly recognized risks. The Food and Drug Administration (FDA) in the United States banned feeding cattle body parts to other cows in 1997; however, this new restriction came with many exemptions. Feeding blood products and fat is still permitted. It is also still permissible to feed nonruminant animals (e.g., chickens, pigs, fish) and their litter to cattle, as well as the reverse: feeding cattle to chicken, pigs, and fish.[70] Cattle remains can also be fed to other cows via waste products, such as those obtained from restaurants.[71] Further, in the wake of a massive pet food safety crisis in Canada and the United States in 2007, where millions of pounds of pet food was recalled because it was contaminated with melamine that was sickening and killing companion animals, it was also revealed that livestock animals were also being fed pet food deemed unfit for consumption by pets. Some of the livestock animals who had consumed the contaminated pet food were subsequently slaughtered and entered the human food supply; many others were killed as a preventable measure.[72] Even with all of these loopholes in the regulations, workers in the industry report that items that were actually banned in 1997 are still being put into the food supply, and regulators have also uncovered violations.[73] In 2004 alone, the FDA found nearly one hundred companies were violating the feed regulations.[74]

Shukin draws an interesting parallel between the human variant of mad cow disease attacking the human brain and the 9/11 terrorist attack against New York, the epicenter of the global economy; she asserts both are materially symptomatic of neoliberal culture. Her analogy is thought-provoking, and to it I would add that in both cases the demonstrative governmental response made citizens feel safer but interrogating the root causes of the incidents was seemingly taboo. People did not, and still generally do not, want to know what livestock animals are fed to make them grow so large, so fast.[75]

Influenza

When we think of illnesses that can be contracted from meat, the two just examined—bacteriological contamination and mad cow disease—spring rather easily to mind. Another class of illnesses, however, are also connected to the way animals used for meat are currently produced and may pose even greater future risks than the illnesses just discussed, yet they are not publicly linked to animal agriculture as frequently and thoroughly. The conditions in CAFOs—large numbers of animals, in close contact, and stressed—breed viruses, which spread rapidly among the animals. Whereas viruses introduced into areas where humans are confined in relatively small spaces, such as military bases, work their way through the population and run out of new hosts, in CAFOs, new and previously unexposed animals are continually added, so the viruses continue to mutate and spread. They are transmitted by water, air, and physical touch and can mutate and spread to other organisms in CAFOs (e.g., flies, rats, and humans) and beyond.[76]

There is growing concern about these zoonotic (meaning that they can cross species lines) influenza strains and speculation that the next human pandemic will originate in animals and breach the species line with deadly consequences.[77] Recent years have witnessed the emergence of rather virulent strains of zoonotic influenza, commonly referred to as the *bird flu* and *swine flu*. The bird flu was first detected in poultry CAFOs in Asia in the 1990s.[78] In 2004, a strain of this flu spread around Asia; thirty-four people became ill and twenty-three died. Additionally, approximately two hundred million birds were killed in an attempt to stem the transmission.[79] Five years later, H1N1, a virus commonly occurring in pigs, breached the species line and reached the status of a global human pandemic.[80] The CDC estimates that between April 2009 and April 2010, there were between forty-three and eighty-nine million cases of human H1N1 infection. Between 195,000 and 403,000 people were hospitalized as a result, and there were between 8,870 and 18,300 related deaths.[81]

Research into H1N1 transmission has found that CAFO workers are at increased risk of contracting it. Authors of one study concluded that the chance of finding H1N1 antibodies in CAFO workers' systems is fifty times greater than the likelihood of finding them in a control group. Additionally, their spouses were twenty-five times more likely to have the antibodies in their systems.[82] Thus, it appears that not only are CAFO workers at greater risk, but the people they come into contact with are as well. Saenz and colleagues developed a mathematical model of influenza transmission to map transmission from CAFO workers and as a result conclude that "the presence of CAFO workers increases dramatically the size of the epidemic and that these effects are greater as the percentage of CAFO workers [in a community] increases."[83] More specifically, when 15 percent of a community is CAFO workers, the proportion of infected persons is estimated at 42 percent. This proportion increases to 86 percent when CAFO workers comprise 45 percent of the population.[84] The industry is currently not required to test for these viruses because they are not on the list of illnesses that have to be reported. Despite the real risk to their workers and the general population, the industry has resisted the implementation of processes to detect such viruses among their animals, just as they did with bacteriological testing. The pork industry in particular has reportedly been keeping public health officials out of their farms because they fear the economic consequences of positive virus detection.[85]

For their part, the scientific community has been relatively quiet about the potential risks posed by these flu viruses and what the industry could do to be proactive. A senior officer with the Pew Environmental Group who has voiced concern about the inaction of regulators and scientists on this issue, speculates

> that competing financial interests may be partly to blame for the current lack of data and regulation. "Even the best scientists seem loathe to say anything against the industry," he says. "With the decline in public research funding, it's industrial animal agriculture that pays for virtually all the animal sciences research going on at land-grant universities today."[86]

Of course, the problems associated with getting research funded are not specific to the issue of influenza; to varying degrees these problems complicate the assessment of and interventions in all of the impacts of industrial animal agriculture discussed in this chapter. In spite of these challenges, the scientific community has begun to voice concerns about another issue arising in the industry: the widespread use of antibiotics.

Antibiotic Use and Resistance

The bacteria, prions, and viruses facilitated by industrial animal agriculture are not the only potential risks to human health emerging therefrom. Industrial animal agriculture uses significant amounts of antibiotics. The problem with the heavy use of antibiotics in the industry is that the bacteria that survive the antibiotics proceed to multiply, thereby creating a line of antibiotic resistant bacteria, such as *Staphylococcus aureus* (Staph).[87] People can therefore be harmed indirectly by the practices of industrial animal agriculture when the antibiotics prescribed to treat them when they are ill do not work.

As mentioned above, the conditions inside CAFOs create a perfect storm for illnesses among livestock animals. Beginning in the 1950s, the industry turned to the use of antibiotics to treat and prevent some of these illnesses. Antibiotics also had the added benefit of promoting growth in livestock animals (like adding the remains and litter of other animals to their feed). Since the 1950s, the amount of antibiotics used in the industry has increased precipitously, and the amount used today is no less than astounding. The Union of Concerned Scientists estimates that as much as 87 percent of antibiotics currently used in the United States are used on animals.[88] Globally, the proportion used on animals is approximately 50 percent.[89] Given the trend toward industrializing animal agriculture in developing countries, we can expect the proportion used on animals globally will increase.

Antibiotic-resistant bacteria were first discovered in the mid-1970s on chicken farms where there was widespread use of antibiotics.[90] Recent samples taken in a swine CAFO found 98 percent of the isolates tested were resistant to at least two antibiotics frequently used to promote pig growth. The antibiotics and the resistant bacteria, however, extend beyond the confines of farms. Estimates indicate that as much as 75 percent of antibiotics given to farm animals pass through the animal's body and into the wider environment.[91] Studies have demonstrated that CAFOs and the areas surrounding them are hotbeds for antibiotic-resistant bacteria.[92] The public can also be exposed to antibiotic-resistant bacteria in the community and by handling meat or consuming meat that has not been cooked enough to kill the bacteria. The presence of antibiotic resistant bacteria in meat is not uncommon. An analysis of two hundred specimens of ground meat for sale in Washington, D.C., found that 20 percent were contaminated with salmonella and 84 percent were resistant to one antibiotic, while 53 percent were resistant to at least three.[93]

Currently in the United States, antibiotic-resistant infections are responsible for ninety thousand deaths annually.[94] Methicillin-resistant *Staphylococcus aureus* (MRSA), which is known to have originated on pig farms, is now responsible for more deaths per year in the United States than HIV/AIDS.[95] This resistance is not only a threat to human health, it is extremely taxing on the health care system, costing approximately $30 million per year.[96] And the situation is getting worse instead of better. Today, "antibiotic resistance is increasing among most human pathogens. The many bacteria resistant to multiple antibiotics in particular has heightened concern. In some cases there are few or no antibiotics available to treat resistant pathogens."[97] The current levels of antibiotic resistance are so significant that it has been suggested that we are on the verge of a postantibiotic era. In this era, treatments for many life-threatening infections would be ineffective, and many more deaths would result.

The growing danger posed by antibiotic-resistant organisms is receiving increasing international attention. Several countries, including Sweden, Denmark, and members of the European Union, have banned subtherapeutic antibiotic use in livestock. Subtherapeutic use is not intended to treat medical conditions; instead, it is intended to prevent illness and increase growth rates among the treated animals. These bans appear to have had the desired effect. For instance, research in Denmark documented reduced antibiotic-resistance after subtherapeutic use was discontinued there. In light of such studies, the World Health Organization has called for a phasing-out of the use of antibiotics for growth promotion in animal agriculture. It also estimates that the costs of eliminating the subtherapeutic use of antibiotics in the industry are relatively modest; the estimated cost increase for raising pigs without the use of antibiotics, for instance, is only 1 percent.[98]

There have also been calls here for discontinuing the subtherapeutic use of antibiotics in animal agriculture. In 2001, the American Medical Association passed a resolution opposing the subtherapeutic use of antibiotics in animals.[99] The industry, however, has thus far failed to acknowledge the problems associated with this practice and has even been accused of burying science on the subject. In one case, a microbiologist resigned from the USDA because on at least ten occasions he was prevented from speaking about his publicly funded research on antibiotic-resistant bacteria in CAFOs, reportedly due to pressure exerted on the USDA by the animal agriculture industry.[100]

ENVIRONMENTAL IMPACTS

Studies of the environmental impacts of the industry expose numerous other externalities, many of which also pose risks to human health, that the industry would no doubt prefer remain unaccounted for. This research indicates that the impacts are numerous and significant. The scope and significance of these impacts is summed up by the Worldwatch Institute as follows:

The human appetite for animal flesh is a driving force behind virtually every major category of environmental damage now threatening the human future—deforestation, erosion, fresh water scarcity, air and water pollution, climate change, biodiversity loss, social injustice, the destabilization of communities and the spread of disease.[101]

Space constraints preclude being able to give sufficient attention to all of these environmental impacts here; instead I focus on two of the most pressing issues: pollution and global climate change. The research discussed in this section therefore provides a partial picture of the "ecological hoofprint" of the industry.[102]

Pollution

Industrial animal agriculture is a major polluter, and Hahn Niman claims it to be second only to the coal industry in getting away with systematic pollution. Most of the pollution occurs in the production segment of the industry. The primary forms of pollution the industry contributes to are air and water, so that is where attention is focused here.

Air Pollution

The air pollution caused by livestock production is concentrated in and around CAFOs, where the air has been found to contain elevated levels of bioaerosols and endotoxins.[103] Recent research has documented significant negative human health effects of this air pollution, particularly among those working within it. CAFO workers are more likely than the general public to suffer from respiratory problems, which is not all that surprising given their proximity to the source of pollutants.[104] The air around manure lagoons can be particularly dangerous. The fumes in the lagoons are so strong that workers have succumbed to them and died. For instance, a man repairing a lagoon in Michigan fell in upon being overcome by the fumes. He died, along with his nephew, cousin, brother, and father who jumped in to try and save him.[105]

The negative impacts of air pollution also extend beyond the CAFOs and into the larger community. Increased cases of respiratory illnesses, digestive problems, mood disorders, and sleep disorders, all attributable to the compounds in the air, have been documented in CAFO communities.[106] Bacteria dispersion via the air is also a problem. A study of bacteria upwind and downwind of swine CAFOs found that the amount of bacteria within CAFOs and downwind of them were high enough to be considered a human health hazard. As a result of these findings, the researchers recommend siting CAFOs at least two hundred meters away from residential areas.[107]

Further, research indicates that it is not just working in or living near a CAFO that puts one at risk for developing health problems, but even attending a school near a CAFO is associated with negative respiratory effects. Mirabelli and colleagues analyzed the relationship between CAFO exposure and asthma symptoms in a sample of sixty thousand children between the ages of twelve and fourteen years in North Carolina. They found that wheezing within the past year was 5 percent more likely if a participant attended a school within three miles of a CAFO and 24 percent more likely if odors from the CAFO were noticeable at the school at least twice a month. The authors explain that the CAFO effects they document in the study are of comparable strength as other risk factors for respiratory problems, such as age, race, wealth, use of a gas stove, and exposure to secondhand cigarette smoke.

To date, regulation of air pollution from CAFOs via the Clean Air Act has not been what one would define as successful. Enforcement has been limited, and when violations are detected they are not necessarily acted upon. Most notably, in 2005, the EPA excused the violations of 6,700 CAFOs with the stated goal of measuring air emissions. The pollution recorded was significant enough that some of the excused CAFOs would have been responsible for paying fees of upward of $27,000 per day.[108]

Water Pollution

Agriculture is also consistently implicated as the leading cause of pollution in rivers, streams, reservoirs, and lakes.[109] In order to understand why this is the case it is necessary to once again examine the rather undesirable topic of manure. Research indicates that, like the air inside and around CAFOs, the manure from the animals can contain a variety of toxic substances, including ammonia, methane, hydrogen sulfide, carbon monoxide, cyanide, phosphorous, nitrates, disinfectants, and heavy metals. It can also contain remnants of drugs given to the animals and more than one hundred types of pathogens.[110]

Not only does manure contain a number of potentially dangerous substances, the industry produces A LOT of it. A CAFO containing five hundred thousand pigs produces more waste than that generated by the 1.5 million human inhabitants of Manhattan.[111] The General Accounting Office estimates that nationally the amount of animal waste produced is 130 times the amount produced by people; it adds up to approximately five tons of animal waste produced for every citizen.[112] In total, an estimated five hundred million tons of manure has to be disposed of in the United States each year.[113]

Whereas small farms might be able to use all the manure they produce on their land as a fertilizer, large CAFOs simply produce too much manure to use in this way, and it must instead be stored. Large facilities store the manure in constructed lagoons. These lagoons can be immense. For instance, Smithfield's largest lagoon is 120,000 square feet in size. Manure has been known to breach the confines of these lagoons. Although they are lined, the liners can tear, allowing manure to escape. Light rains have also caused overflows and "major floods have transformed entire counties into pig-shit bayous."[114] North Carolina experienced such a catastrophe in 1995 when a manure lagoon eight acres in size ruptured and resulted in a spill two times the size of the Exxon Valdez spill.[115] There were further problems four years later when Hurricane Floyd facilitated the release of 120 million gallons of manure into local waterways.[116]

It is not just the storage of manure that can be problematic. Although there is too much manure to apply all of it as fertilizer, a portion of it is used in this manner. This can also be harmful. Whereas human sewage must be treated for disposal, animal manure is applied untreated onto crop land.[117] Treating animal manure would be costly, so much so that the industry would lose money if they had to treat manure as human waste is treated.[118] Therefore, it continues to be applied untreated.

When untreated manure leaks out of lagoons or is applied to the land it can seep into surface water, bringing contaminants, bacteria, and high levels of nutrients with it. The additional nutrients, such as nitrogen and phosphorus, cause eutrophication, which depletes oxygen and kills fish, and consequently also negatively affects animal populations that rely on fish consumption. This harmful process is facilitated by rain, yet the industry argues that this runoff is exempt from regulation.[119]

Research has documented significant water quality problems near CAFOs. The Department of Environmental Quality took weekly water samples around a CAFO in Michigan and found that 98 percent of them had E. coli levels that exceeded the water quality standard.[120] Some studies have even found that water quality near CAFOs is worse than that downstream of wastewater treatment plants. For instance, West and colleagues found in their study that water quality downstream of wastewater treatment plants was markedly better than around CAFOs. They also found the sites near CAFOs had levels of phosphorus one to eight times higher than the control sites, and all of the CAFO sites tested exceeded the threshold of acceptable phosphorus levels. Additionally, 41.6 percent of the CAFO sites contained bacteria resistant to multiple antibiotics, compared to 16.5 percent of the other sites examined.[121]

These problems persist under existing laws and levels of regulatory enforcement. While some companies have been charged with offences under the Clean Water Act, the penalties may not be significant enough to produce a deterrent effect. For instance, in 1998, Smithfield was charged with 6,900 violations of the Clean Water Act and fined $12.6 million.[122] Although this seems like an impressive fine, it only actually amounted to 0.035 percent of the company's annual revenue.[123]

Global Climate Change

Global climate change has gained prominence as *the* environmental issue of the twenty-first century. The general public and social institutions have become increasingly concerned about the causes of global climate change and ways to mitigate its effects. Yet a major contributor to global climate change—industrial animal agriculture—has flown largely under the radar. A report published by the UN's Food and Agriculture Organization (FAO) in 2006, however, began to put industrial animal agriculture front and center.

The report, entitled "Livestock's Long Shadow," details research conducted to assess the specific contribution of animal agriculture to global climate change.[124] To do so, the researchers aggregated the emissions produced along the livestock commodity chain. Their analysis determined that 18 percent of total global greenhouse gas emissions are produced by animal agriculture—a greater proportion than that produced by the transportation sector, which is interesting because so much of the attention paid to mitigating global climate change has been focused on changing public and private transportation strategies. World Bank researchers have since suggested that the contribution of animal agriculture to global climate change estimated by the FAO is actually too low. They instead estimate that animal agriculture is responsible for 51 percent of greenhouse gas emissions.[125] To put these estimates into perspective, an average U.S. consumer reducing his/her meat consumption by one-fifth reduces the amount of greenhouse gases emitted roughly as much as switching to a hybrid vehicle.[126]

A detailed breakdown of the anthropogenic production of greenhouse gases reveals that animal agriculture is responsible for 9 percent of carbon dioxide, 37 percent of methane, and 65 percent of nitrous oxide emissions. The industry contributes carbon dioxide in two major ways. The first is through deforestation.[127] Deforestation for meat production occurs when forests are destroyed to make room for pastures (evident in Central America) or to raise the crops that will be converted to feed (evident in South America). Deforestation not only produces excess carbon dioxide, it also removes carbon sinks.[128] Deforestation tied to the industry therefore upsets the carbon cycle and is responsible for approximately 2.4 billion tons of carbon dioxide a year. The second way animal agriculture contributes to the production of carbon dioxide is through the use of fossil fuels in the production of livestock feed, as well as through transporting supplies and products (e.g., feed, animals to slaughter). The digestion of feed and emission of livestock waste produces methane gas. Finally, the industry contributes to nitrous oxide via the production of leguminous feed crops and the application of chemical fertilizers to those crops.[129]

Despite the significant contributions of animal agriculture to problematic greenhouse gases detailed in the "Livestock's Long Shadow" report, an analysis of animal advocacy organizations, environmental organizations, meat industry stakeholder groups, governmental agencies, and newspapers in the United States and Canada reveals that discourses acknowledging the link between animal agriculture and global climate change have not increased since the release of the report. The study also finds that while animal advocacy organizations have been articulating the connection between animal agriculture and global climate change, environmental organizations and the media have been relatively silent on the issue, while the industry stakeholder groups and governmental agencies are downplaying the connection.[130]

The contribution of industrial animal agriculture to greenhouse gases can be expected to increase because, as discussed earlier in the book, developing countries are adopting more meat-focused diets and turning to industrial production methods in order to meet the increased demand. Not only will the global increase in meat consumption and production likely exacerbate global climate change and the other environmental, animal welfare, and human health problems discussed earlier, it will create food security challenges, contributing to what is referred to as food *in*security.

FOOD INSECURITY

The notion that increasing meat production could contribute to food insecurity may seem counterintuitive; however, it could very well do so in two main ways. First of all, producing more meat will require increasing livestock production, which will by extension necessitate increasing feed production because current global feed production levels will not be able to sustain higher levels of meat production.[131] It will therefore be necessary to increase feed production, which can only be accomplished in three ways: by reallocating land currently used for other purposes (e.g., housing) to crop production, reallocating cropland currently used for production of subsistence crops for human consumption to crops for livestock feed, and/or intensifying the production of crops on the land that is available to produce both feed and subsistence crops. Due to natural limits, the latter option is limited and will not be able to increase production enough. Therefore, it is inevitable that some land currently used for other purposes (e.g., subsistence crops, housing) will be reallocated to produce feed for livestock. This process has already begun,[132] and it "has further compounded their [developing countries'] neocolonial status."[133] In addition to clearing forests, the spread of cattle production has displaced already disadvantaged human populations and small farmers and increased the concentration of land ownership."[134]

As this process unfolds and results in the displacement of people, as well as the further concentration of power, it will also contribute to food insecurities as the production of crops shifts away from feeding people directly to feeding livestock animals, which are then killed and fed to people. Meat-based diets are notoriously inefficient; for example, it takes eleven to seventeen calories of feed to produce one calorie of beef, and one acre used for meat production produces one fifth as much protein as an acre used to produce cereal crops does.[135] Prioritizing feeding grain to livestock instead of people is already contributing to global malnutrition and famine.[136] In particular, the export of grain produced in developing countries for use as animal feed instead of using it to feed domestic citizens has been going on for years. In fact, this process contributed to the massive famine in Ethiopia in the 1980s.[137]

The second way that the expansion of meat consumption in developing countries will contribute to food insecurity has to do more directly with importing. Countries participating in the General Agreement on Tariffs and Trade and the World Trade Organization are extremely limited in their ability to impose tariffs on imported products. Therefore, some countries are compelled to allow the import of cheaper, industrially produced meat, which makes it more difficult for the local industry to compete, and in some instances the local industry withers away.[138]

The loss of local industry compromises the control localities have over their food production, thereby contributing to food insecurity. The loss of local food production capacity and the loss of land to animal feed production, coupled with the dual pressures of an increasing human population and increasing levels of per capita meat consumption, will have the perhaps unintended consequences of exposing an increasing number of people globally to malnutrition and starvation.[139] Therefore, although it is often assumed that increasing meat production will lead to increased food security, it is more complicated than that. Likewise, the notion that growth in industrial animal agriculture and processing in a locality results in increased quality employment opportunities is not necessarily borne out.

COMMUNITY IMPACTS

While the plight of individual workers within the industry detailed earlier has not received the amount of attention it deserves, the effects on communities have received even less. This dearth of attention prompted Marcus to point out in his book examining changes in the industry since the mid-twentieth century, "while stories of work-related tragedies at slaughterhouses are commonplace, the impact that the facilities have on communities is every bit as disturbing."[140] It is generally assumed that opening animal production facilities (i.e., CAFOs) and processing facilities (i.e., slaughterhouses) will benefit communities. The expectation is that such operations will bring increased jobs and economic prosperity to communities. On the face of it, that assumption seems reasonable; however, further examination reveals that it does not necessarily apply to industrial animal agriculture.

The concentrated and consolidated animal agriculture industry of today does not contribute to the community like it did when the industry was more decentralized. Recall Goldschmidt's research described earlier in the book. He found in the mid-twentieth century that communities with small-scale farms performed more positively on economic and social measures than did communities with the large industrial facilities.[141] More recent research on the economic impacts of the animal processing industry finds that, although these facilities are associated with some county employment growth, they are also associated with lowered wage growth and decreased growth in other areas of the economy.[142] Due to the high degree of vertical integration in the industry, detailed earlier, animal agriculture companies generally do not purchase the supplies they need from the local community; instead they purchase from companies within the same corporate family, often headquartered elsewhere. Therefore, industrial animal agriculture does not benefit communities in the ways that some other industries do. In fact, the Farmers Union has testified in the Senate that the industry is actually harming communities, or as they put it, "sucking the lifeblood out of rural communities."[143]

Although the industry does contribute to the local tax base, the taxes these companies pay are often outweighed by the costs of the subsidies they are given and the externalities they produce. Municipal, state, and federal tax money is used for building CAFOs, pollution control, subsidizing feed and energy costs, and funding agricultural research. The Union of Concerned Scientists has estimated that at least $7 billion in tax money per year goes to CAFOs and an additional $4.1 billion per year is directed specifically toward trying to control manure leaking from these facilities. The additional indirect costs of environmental and public health impacts that the industry has externalized, discussed earlier, are more difficult to quantify.[144]

One impact that has been reliably quantified is the effect on property values. Researchers estimate that vacant land within three miles of a CAFO in Missouri experiences a 6.6 percent decline in value, and a residence within one-tenth of a mile of a CAFO loses approximately 88.3 percent in value. Other studies in Washington and Michigan estimate 50 percent value loses of property proximate to CAFOs.[145] The total decline in property values related to the industry in the United States is estimated at $26 billion.[146]

Taking all of these variables into consideration, it is apparent that the notion that industrial animal agriculture brings prosperity to communities in the form of more jobs, tax revenue, and local market stimulation is short sighted. The costs of subsidies and externalities must be entered into the equation. Noble articulates a more holistic understanding of this equation as follows:

> CAFOs can violate environmental laws with impunity and routinely impose public health hazards on surrounding populations, while neighboring residents and communities must bear the costs of coping with the pollution or pony up with the rest of us to pay the polluters, even for measures that actually increase the CAFO fecal flood. Meanwhile, the factory farm sector trumpets its "economic efficiency" while collecting billions of dollars in public subsidies at all levels of government, and every year livestock and poultry CAFOs grow larger and more concentrated. Sustainable farmers and ranchers are left to compete on an uneven playing field against some of the world's largest corporations and their allies.[147]

The potential negative impacts of the industry on people in those communities are not just economic. There can also be negative health impacts, detailed earlier. To summarize, residents living near CAFOs experience greater rates of headache, runny nose, sore throat, excessive coughing, diarrhea, and burning eyes than residents who do not live near CAFOs. Even for those who are not sickened, quality of life can be compromised: over half of those living near CAFOs report that they cannot open windows or spend time outside on nice days.[148]

Animal production facilities can also have disruptive social effects. The siting of these facilities has been known to create conflict between community members. Additionally, it has been reported that those living amid CAFOs even exhibit increased post-traumatic stress disorder (PTSD) symptoms.[149] These effects may be disproportionately foisted upon already vulnerable communities, as researchers studying the siting of pig CAFOs, particularly corporate-owned facilities, have concluded that they tend to be located in low-income, minority neighborhoods.[150]

It is not only the animal production segment of the industry that can negatively affect communities; the animal slaughtering and processing segment can have different negative impacts on communities. In *The Jungle*, Upton Sinclair detailed not only the problems taking place inside the slaughterhouses in Chicago at the turn of the twentieth century, but also problems in the broader community that seemed to be emanating from the slaughterhouses. When many of the large slaughterhouses, such as those Sinclair wrote about, moved from large urban centers to small towns in the second half of the twentieth century, it created an almost perfect environment for assessing the impacts of slaughterhouses on communities because researchers could analyze pre- and postslaughterhouse effects.

Some of the changes in communities where large slaughterhouses opened, such as housing shortages, could be explained by the fundamental increase in the number of people in these communities and were not unique to the animal processing industry: any other industry that moved into these small communities and drew thousands of employees would have the same effect on housing. The new slaughterhouse communities also experienced increased demands for social assistance.[151] Generally, we would not expect that a new facility that employs hundreds or even thousands would be associated with an increase in social assistance demands. Here the unique effects of the animal processing industry on communities start to become visible. The characteristics of slaughterhouse work—relatively low wages, high injury and illness rates, and high employee turnover rates—combine to create a greater need for social assistance among those who are attracted to these communities for work and then find themselves chewed up and spit out by the industry.

Ethnographic studies in new slaughterhouse communities (such as Finney County, Kansas; Lexington, Nebraska; Perry and Storm Lake, Iowa; Guymon, Oklahoma; and Brooks, Alberta) documented another social change: crime rates increased after the slaughterhouses came to town. For instance, in the five-year period following the opening of two slaughterhouses, Finney County, Kansas, witnessed a 130 percent increase violent crimes.[152] In addition to increases in violent crime, increases in property crimes and drug offences were also observed.[153] An explanation for this phenomenon was not immediately obvious.[154]

A number of theories to explain the crime rate increases were proposed. The first group of theories centered on the demographics of the animal slaughtering workforce. The general public, media, government officials, and industry spokespeople pointed to the reliance of the industry on immigrant labor and the resultant influx of immigrants into these communities as an explanation for the increases in crime.[155] Academics focused on other demographic factors, specifically the fact that the industry attracted young single males, a group that poses the greatest criminogenic risk more generally.[156]

The second group of theories suggested that the increases in crime rates could be explained by the population booms experienced by these small towns and resultant social disorganization. This theorizing suggests that preboom communities are stable and cohesive. An influx of a large number of newcomers can disrupt the stability and reduce the homogeneity of the community, which in turn can reduce the efficacy of informal social control and contribute to deviance.[157]

It was also hypothesized in the literature that the high employee turnover rates in the industry could contribute to increased unemployment in slaughterhouse communities because those who came to the community to work in the slaughterhouse may then find themselves unemployed.[158] Unemployment could be a tipping point for former slaughterhouse workers and a factor pushing individuals toward criminal offending.

None of these theories, however, takes into consideration the unique type of work undertaken inside slaughterhouses: in these workplaces the materials being transformed are not inanimate objects—they are live, dying, and dead animals. My coauthors Linda Kalof, Thomas Dietz, and I coined the term the "Sinclair hypothesis" to refer to the potential spillover of the violence inherent in the industry outside to the larger community,[159] because in *The Jungle*, Sinclair posits that the violence he observed in the communities around the Chicago slaughterhouses had originated in and spilled over from the slaughterhouses, where violence is a necessary part of turning live animals into meat. His observations include the following:

> He [the police officer] has to be prompt—for these two-o'clock-in-the-morning fights, if they once get out of hand, are like a forest fire, and may mean the whole reserves at the station. The thing to do is to crack every fighting head that you can see, before there are so many fighting heads that you cannot crack any of them. There is but scant account kept of cracked heads in back of the [stock] yards, *for men who have to crack the heads of animals all day seem to get into the habit, and to practice on their friends, and even on their families, between times.*[160]

We conducted a study to examine the potential of the Sinclair hypothesis by assessing the relationship between slaughterhouse employment levels and crime in counties in the United States. We analyzed crime rates in 581 nonmetropolitan counties between 1994 and 2002, statistically controlling for common correlates of crime, such as gender, age, income, population influxes, unemployment, and population density. We found that slaughterhouse employment levels in counties are related to increases in total arrest rates, as well as arrests for violent crimes, rape, and other sex offences. We also included a number of comparison industries in our analyses that are similar to slaughterhouse work in the proportion of immigrants in the workforce and injuries and illnesses on the job. Overall, we found that the relationship between slaughterhouse employment levels and crime was unique compared with these comparison industries; in fact, many of the comparison industries were instead negatively related to crime rates.[161]

Another study set out to assess the economic impacts of the animal processing industry on local economies and included crime rates in their analyses. The researchers examined 1,404 U.S. counties from 1990 to 2000. When they examined the effect of the meatpacking/slaughtering and processing industries combined, no significant changes in property and violent crimes were observed between counties with and without the industries. When they examined the meatpacking or slaughterhouse industry separately, however, they found "counties with growth in meat packing . . . experienced faster growth in violent crime rates over the decade relative to counties without packing plants."[162]

These quantitative studies support and expand upon the observations of crime increases in the earlier ethnographic studies of individual slaughterhouse communities: they indicate that at the county level at least, there is a relationship between slaughterhouse employment levels and crime rates, particularly violent crimes, even when controlling for a number of possible intervening variables. Further research is needed, however, to understand the mechanisms involved in these relationships. Due to the aggregate level of analysis in these studies it is not possible to assess who exactly is committing these crimes and why, and if it is indeed slaughterhouse workers who are behind the increases in crime, and if so, it is unclear if this offending is new to them upon working in the industry or if the crime increases merely reflect that they are new to the community. Future research assessing these possibilities might usefully draw theoretically and methodologically from previous studies examining the spillover of violence from workplaces to communities, such as among military personnel and prison guards.[163] Future research might also draw on speculation by some criminologists that the sanctioning of violence can extend beyond local communities and have more of a widespread cultural impact. For instance, the "brutalization hypothesis" asserts that the state's use of the death penalty legitimizes the use of lethal violence instead of deterring its use.[164] This theorizing raises an interesting question in the context of industrial animal agriculture: Is it possible that the growth of industrialized animal agriculture and the reduction of animals to objects therein is having a broader, cultural impact? And might it exacerbate a growing cultural tension between our use of some animals as food and our protective affection for others?

CULTURAL IMPACTS

The growing cultural tension brewing in the United States is the result of two competing impulses. On the one hand, as discussed throughout this book, the primary objective of industrial animal agriculture is to generate profits. Given this mandate, industrial agribusinesses raise and kill animals in the most cost-efficient ways, which are often at odds with animal well-being. On the other hand, there is growing interest in and affection for some species of animals among the general public, and there is a possibility that this interest and affection could eventually spill over to the species of animals currently used by industrial animal agriculture. Some animal advocacy organizations are working specifically to facilitate this spillover in the hope that the general public demands changes to the way livestock animals are raised and killed in response. In order to appreciate the transformative possibilities of this tension, it is necessary to first explore its origins.

During the Enlightenment period, the primacy of religion in society was challenged, and reason was championed as an alternative. While the questioning of religion and superstition at this time was not in and of itself a bad thing as far as the treatment of animals was concerned, the growing influence of rationality and science turned out to have negative consequences of their own. The new ideologies of the day contributed to the objectification of nature. As a product of the larger culture, agriculture began to adopt the new "rational" ideologies, and its goals shifted accordingly.[165]

In his examination of the ideologies undergirding modern agriculture, Kimbrell identifies objectivity, efficiency, competition, and progress as being particularly influential.[166] Adherence to these ideologies had significant consequences. He explains that the new "cult of objectivity" bracketed out subjectivity, and those who challenged the sanctity of objectivity were (and often still are) denigrated as being irrational and emotional (read feminine). Efficiency also became culturally revered, particularly as a means to actualize the goal of being more competitive. The shifting goals of agriculture were viewed as being part of a larger march toward progress, which took on an almost religious character itself, discussed earlier in this book. Framing current developments in industrial animal agriculture under this banner of progress has many consequences, but one particularly problematic one is that critiquing the industry has come to be viewed as backward or even deviant, and actions are being taken to curtail such critiques.

Power imbalances are also making sustained critiques difficult. Reflecting upon the power imbalance between industrial animal agriculture, its critics, and the general public, Thu and Joy argue the industry is engaged in eroding the democratic process; Joy asserts that "our democracy has become a meatocracy."[167] This meatocracy comes complete with laws banning the disparagement of agriculture that threaten freedom of speech, and ag-gag laws banning photographing and filming CAFOs that threaten whistle-blower protections. Thu concludes that the erosion of citizen rights by the industry makes industrial animal agriculture an issue that everyone ought to be concerned with, for as he explains it, "CAFOs are in everyone's backyard."[168] Within this larger cultural context, corporate decision-makers who put profits ahead of the well-being of animals, workers, and citizens are not deviant individuals. Instead, they are acting in accordance with and reinforcing current cultural norms, which are supported by social institutions, including "law and markets [that] direct corporate managers, particularly managers of publicly traded corporations, to make the most productive use of their animal property to maximize profits."[169]

Not only is there ideological support for industrial animal production and processing, but the consumption of animals is also culturally supported and normalized. In an examination of materials promoting the meat industry and their products (i.e., advertisements, industry websites), Nik Taylor and I found that there is a hegemonic discourse advocating not only the acceptability of eating animals, but also the necessity of doing so. Based on our analyses, we argue this discourse is supported by narratives of "happy" animals who give their lives up so that we can consume them, the alleged "naturalness" of current meat products and of meat consumption more generally, and the healthfulness of consuming meat products.[170]

While various ideologies rationalize the continued, and increasing, production, processing, and consumption of animals on the one hand, the seeds that could mature into a cultural chasm are taking root on the other. Recall historian Richard Bulliet's conceptual division of our relationships with animals into two major eras: domesticity and postdomesticity. No doubt this oversimplifies a more complicated history; however, it usefully elucidates growing tensions in our relationships with animals. Bulliet asserts that, in the domestic era, social and economic structures made it possible for people to come into regular contact with animals, and not just "pet" species. The subsequent postdomestic era, which Bulliet asserts emerged in the 1970s, is characterized instead by a physical and psychological distancing between people and the animals who are used to produce products for human use.[171] I would suggest that this aspect of postdomesticity began to take shape with the industrialization of animal agriculture around the turn of the twentieth century. Regardless, another aspect of the postdomestic era, which did seem to become more widespread in the 1970s, is that the majority of the population is cultivating very close relationships with species of animals kept as pets.

The tension I referenced earlier develops in the postdomestic era because

> a postdomestic society emerging from domestic antecedents continues to consume animal products in abundance, but psychologically, its members experience feelings of guilt, shame, and disgust when they think (as seldom as possible) about the industrial processes by which domestic animals are rendered into products and about how those products come to market.[172]

This tension is exacerbated by the fact that the number of animals being used for human consumption is increasing at the same time that their quality of life is arguably diminishing. Even if there was a fully successful way to block all of this out, Otter warns that "institutionalized forgetting might create the conditions of possibility for cruelty of a new kind, on a greater, more deeply hidden scale."[173]

Melanie Joy argues that the cultural attempt at forgetting is an element of all violent ideologies. She refers to it as "knowing without knowing," akin to what philosopher Nancy Williams refers to as affected ignorance.[174] Joy asserts that this knowing without knowing is an essential part of what she refers to as *carnism*, an ideology that normalizes meat consumption. This ideology is pervasive and influences our beliefs and attitudes about animals and our consumption of them. Like patriarchy, for instance, this ideology simultaneously normalizes and justifies the subjugation of another group. In the case of patriarchy, it results in sexism; within carnism, the result is speciesism.

The cultural blinders employed within carnism make it possible to eat some species of animals while sharing our homes, lives, and affection with others. However, like any ideology, it is not totalizing and there are points of rupture and resistance. The resultant fault lines within our carnist culture may be exacerbated by the growing postdomestic mentality Bulliet refers to, as well as the growing evidence of the negative impacts of industry production, processing, and consumption discussed in this chapter: namely, the deterioration of animal welfare, compromised worker safety and well-being, negative human health impacts, negative environmental impacts, growing food insecurity, and negative community and cultural impacts.[175] The possibilities for change brought forth by these fault lines are discussed in the next chapter.

Bridging the Divide between Production, Processing, Consumption, and Impacts

UP TO THIS POINT, THIS BOOK HAS BEEN LARGELY DESCRIPTIVE—PROVIDING INFOR-mation about the production, processing, and consumption of animals for food. This chapter moves away from description and closer to prescription through an examination of alternatives to the current system and what individuals, the state, and companies can do to mitigate the negative impacts of the industry. Many shy away from prescriptive suggestions for a variety of reasons, one of which is that conceptualizing alternatives is difficult work, and with agriculture, the problems are complex and there is not one simple solution.[1] Further, the adequacy of proposed alternatives often generates acrimonious debate, which will also be addressed in this chapter.

These difficulties should not prevent us from considering alternatives and settling for the status quo. As detailed in the previous chapters, simply put, the status quo is not working. Further, some argue that we have a moral responsibility to consider alternatives to current industrial animal agriculture, and this responsibility outweighs the difficulties. Cassuto, for instance, is absolute in his view that we *have* to focus on changing the industry. He states,

> To argue for the current system's preservation because the alternatives are imperfect is to dismiss a moral imperative in the name of short-term expedience. Normative reconfiguration never happens easily. There is, however, no other option. The posthuman era has dawned. The only question remaining is what we will do about it.[2]

This chapter aims to address exactly that question; however, the goal is not to provide directives. Instead this chapter explores an array of alternatives that could ameliorate the numerous negative impacts of the current system. It is not a simple problem, and there are no simple answers.

Before we begin the journey into analyzing these alternatives, I would like to take this opportunity to remind the reader that industrial animal agriculture has occupied but a small amount of time in human history. It can be easy to lose this perspective, particularly when you cannot personally recall a different way of doing things in your lifetime. Failure to understand and appreciate this history can make the current system seem inevitable and natural. It is neither.

This chapter traces the alternatives roughly corresponding to the order of the negative impacts of industrial animal agriculture described in the previous chapter. It begins with relatively modest recommendations for improving industrial animal agriculture, starting with animal welfare modifications, and follows a general pattern of working through more significant (what some might call "radical"), structural suggestions for change.

IMPROVING ANIMAL WELFARE

The previous chapters detailed the serious animal welfare concerns associated with current industrial animal agriculture. These problems do not need to be reiterated here; instead, Scully's words serve as a potent reminder: "when corporate farmers need barbed wire around their happy, sunny lands, and laws to prohibit outsiders from taking photographs . . . and still other laws to exempt farm animals from the definition of 'animals' as covered in federal and state cruelty statutes, something is amiss."[3] The fact that something is amiss is perhaps less contested than what should be done.

At the root of the debate are different answers to one fundamental question: Is it possible to meaningfully improve the lives of livestock animals within the current system? There are two general schools of thought on this. The first, commonly referred to as the animal welfarist, suggests that it is possible and that the best way to improve the lives of animals is to make modifications to the ways in which they are currently produced and processed. It is argued that this is more practical and realistic than attempting to overthrow the system of industrial animal agriculture. In short, good can be accomplished for the animals currently in the system now, instead of holding out for a different system.

The second school of thought, variously referred to as the animal rightist, abolitionist, or even fundamentalist, advocates for the abolition of the instrumental use of animals more generally, and the termination of industrial animal agriculture specifically.[4] Their proffered solution is that people stop consuming animal products entirely. They assert that improvements in animal welfare practices or standards cannot obviate the suffering animals endure by being imprisoned and killed. What is worse, according to this perspective, is that these minor modifications to the system can serve to perpetuate it because people may be lulled into thinking that the consumption of "humanely" treated animals is not problematic.[5]

These two schools of thought are generally presented as being polar opposites, with little common ground in between. This is an oversimplified dichotomization. Francione points out there are some nuanced, although he argues problematic, gradations between the two.[6] Additionally, Mathey and Leahy contend that the chasm between welfarists and abolitionists may not be as wide as has been assumed when it comes to actualizing their goals because animal welfare improvements will result in increased production costs (although the proportion increase varies dramatically depending on the animal and the welfare modification), which companies will pass on to consumers in the form of increased product costs.[7] They assert that increased costs will lead to decreased consumption, as evidenced in Switzerland and Sweden, where egg consumption decreased when battery cages for egg-laying hens were banned. Therefore, the actualization of welfarist objectives may bring abolitionists closer to their goal as well. The possibility remains, however, that welfare reforms could simply promote consumption among those who can afford it. Without conclusive evidence one way or the other, the debate rages on.

The debate between these two schools of thought continues in the background while the general public's concern about animal welfare has been increasing. The reasons for this increasing concern are numerous. It has been suggested that in developed countries, the growing concern exhibited in the past few decades is largely due to an increase in postmaterialistic values in these countries.[8] A shift toward postmaterialistic values is said to occur when the basic needs of a population are met. No longer preoccupied with survival, the population can turn its attention to other concerns, such as quality of life, including the quality of life of others. While this may account somewhat for the increase in concern about animal welfare, I would suggest that there are three other factors that have also come into play: the increasingly disturbing interventions by industrial animal agriculture into the lives of livestock animals, the growing number of people developing close relationships with companion animal species (in what Bulliet describes as the postdomestic era), and the crisis in human health and resultant reflexivity around food consumption that has primed people to begin questioning exactly what they are consuming and how it was produced.[9] These factors have coalesced and contributed to an increase in awareness and concern about animal welfare in industrial animal agriculture.

The degree to which concern for animal welfare has translated into actual reforms has varied across countries. Significant gains have been made in western Europe relative to the United States. I will not delve into great detail about the minutia of the reforms made in Europe, as the focus herein is on the United States, but instead provide a snapshot of the overarching guiding principles and their relevance in the U.S. context.

In 1965, a governmental committee in the United Kingdom investigated farm animal welfare and developed a list of five freedoms intended to guide animal welfare. These five freedoms include freedom from hunger or thirst; freedom from discomfort; freedom from pain, injury, or disease; freedom from fear and distress; and the freedom to express normal behavior. Attention to animal welfare received official recognition in other European countries in 1997, when all EU countries signed on to the Treaty of Amsterdam, which recognizes animals as sentient beings. This treaty places the onus on signatories to take this sentience into consideration when developing animal-related policies. Although this designation is rather abstract, in principle, animal welfare in the European Union is to be measured using multiple indicators, including animal health, productivity, physiology, and ethology. Conversely, in the United States, the well-being of animals is generally assessed only on the basis of health.[10] Therefore, as long as the animal remains physically healthy (i.e., productive) it is assumed that the environment and treatment they are being subjected to are sufficiently humane. Whether or not they are able to engage in their natural behaviors or are stressed as a result of not being able to do so is moot unless it impacts their health.

Although the general principles of animal welfare in the European Union are more progressive than those in the United States, whether they go far enough is a matter of debate. The well-being of animals there, as in the United States, is constrained by the fact that they are still legally considered property. Nonetheless, it is commonly suggested that the United States, which has some of the weakest welfare standards for animals used in agriculture among developed countries,[11] look to Europe as an example and for crafting updated welfare principles. While committing to the five freedoms, recognizing animals as sentient beings, and focusing on more than just the physical health of animals to determine their well-being would certainly improve animal welfare in the United States, doing so is more difficult than simply replicating what has already been done in Europe. Matheny and Leahy point out that there are a number of differences between the U.S. and European contexts that would make simple adoption of European principles in the United States unlikely: animal welfare/rights advocacy in general has a shorter history in the United States than in Europe, and advocates in Europe have historically focused more on animal agriculture than in the United States; the general public in the United States is not as well informed about agricultural practices as European citizens are; there has been greater investment in animal welfare research in Europe; the agriculture lobby and their ability to resist change are stronger in the United States than in Europe; and the industry in Europe does less exporting, which makes it less complicated to regulate animal welfare there.[12]

For these and other reasons, instead of advocating for broad, sweeping changes in animal welfare standards in the United States, similar to those in Europe, it might be more strategically feasible to focus on more minor, incremental changes. There are several relatively easy ways to implement changes that could have large animal welfare payoffs, such as providing downer animals (animals who can no longer walk on their own) with prompt veterinary care or euthanasia as required; replacing the use of carbon dioxide to kill male chicks born into the egg production industry with more humane methods; improving the stunning process for animals prior to slaughter to ensure that they are rendered unconscious; and providing a local anesthetic for calves and pigs before they are castrated.[13] These modest measures would eliminate quite a bit of pain and fear that animals in the industry experience, and the costs of doing so are relatively modest.

Others have suggested a strategy of first focusing welfare improvements on the species that are the most populous in animal agriculture in the United States: poultry and fish. Matheny and Leahy argue that relatively minor changes in the ways these animals are treated would have the largest overall benefit because there are simply so many of them in the industry, and focusing on changes here would have the added benefit of only slightly increasing the cost "per unit" due to the immense scale of the production of these species. They further assert that focusing attention on the animal welfare issues in cattle and pig production could have the unintended impact of actually worsening animal welfare in the aggregate because there is the potential that people may instead consume more fish and poultry due to increased welfare concerns about beef and pork and because of increased beef and pork costs as a result of animal welfare modifications. As evidence of their assertion that focusing critical attention on poultry and fish would have large welfare impacts, Matheny and Leahy calculate that on average if one hundred people eliminated those products from their diets it would positively affect more animals in total than if ninety-nine people went vegan. Put more simply, it might be better for animal welfare in the aggregate to promote the consumption of larger animals instead of many smaller ones, which are raised more intensively.[14] However, there is no assurance that increased demand for pork and beef would not result in worsened conditions for those animals akin to those experienced by poultry and fish.

Employing technology to improve animal welfare is another recommended strategy. Scientists are already at work breeding animals to make them more feed-efficient, produce less harmful fats, and emit less harmful environmental emissions.[15] Those areas of research are being undertaken because they will have economic benefits for the industry. The industry could also use genetic selection to produce animals who have traits that enhance animal welfare, such as lower stress responses. Employing technology in this and other ways to enhance animal welfare is unlikely to generate short-term economic benefits, which makes it less likely that the industry would fund it. It may therefore require investment from other sectors, such as universities and the government.[16] Of course, even if adequate funding could be secured, employing science and technology to improve animal welfare is not a panacea, as it runs the risk of reifying the instrumentalization of animals.[17]

The welfare strategies discussed thus far are directed at the scientific community, industry, and government. Others have suggested more of a bottom-up approach focusing on consumers and the market.[18] They argue that the marketplace is the forum where decisions should be made about animal welfare by producers and consumers through negotiation via supply and demand. The argument is that consumers should vote with their dollars, and if they value animal welfare then companies will respond to that demand. Adams is among those who argue that this is the only possible solution, but he acknowledges that it would require a significant educational component:

> progress can be made toward the lofty goal of minimizing animal suffering only through dramatically increased consumer awareness that will serve to drive market responses. . . . However, as students of economics clearly understand, Adam Smith's "invisible hand" has always been predicated upon the assumption of complete information. Only when consumers are provided sufficient information through the combined efforts of social pressure groups and industry participants, will informed, ethical decision making be possible. To date, that goal seems very distant.[19]

There certainly is a significant knowledge gap between what consumers know about how their food is produced and how they decide to consume. In a 2003 survey of 1,032 likely voters across the United States conducted by Zogby International, the majority of respondents indicated that they are very or somewhat concerned about the treatment of food animals. More than seven in ten respondents also reported that they believe farm animals are treated fairly. The report writers suggest that the fact that a significant majority believe farm animals are treated fairly, while simultaneously claiming to be at least somewhat concerned about the treatment of food animals, indicates that the general public is not receiving enough information about the actual treatment of farm animals and instead their perceptions are being shaped by advertising. This finding may also be reflective of the fact that many people believe that laws are in place to ensure good treatment of farm animals; in the survey, over half of respondents indicated they believe that farm animals are protected by state anticruelty laws and that there are effective state and federal animal welfare laws governing the treatment of farm animals.[20] Other research indicates that, once educated about standard agricultural practices that are legally used and protected, the majority of U.S. citizens object to those practices.[21]

The pressing question then is how can consumers be educated about current industrial animal agriculture practices? Doing so will require a multipronged strategy. One way to educate consumers is through informative academic and nonacademic literature. Unfortunately, many people will not have the time, money, or desire to read such materials. So how can those people be informed?

Social movement organizations, particularly animal advocacy organizations, have been successful in educating at least some consumers about the problems associated with industrial animal agriculture using alternative methods.[22] To date, the degree to which environmental organizations have been involved in educating their constituents about the negative impacts of industrial animal agriculture, environmental and otherwise, has been limited.[23] It should be noted that the organizations that are educating the public about how their food is produced would do well to remember that consumption does not occur in a vacuum: it is impacted by social structural variables (such as gender, race, ethnicity, class, and geographic location), culture, and tradition.[24] These nuances need to be taken into consideration instead of pointing fingers of blame.

The social movement organizations face the same challenge confronted by academics seeking to reach out with educational literature: it can be difficult to get people to pay attention to the message that is being articulated. People are busy, and some simply do not want to know. One strategy that some social movement organizations are using to get around this problem is to pay people for their time: they are paying them to sit and review footage from within the industrial animal agriculture industry so that they can learn where their food comes from. Research on the efficacy of this rather inventive type of approach would prove useful.

Another strategy to educate the public is to lobby for labeling of meat products that provide information about the way the animal(s) used in the products was/were produced and processed, as well as information about the potential health impacts of meat consumption, so that consumers can easily make educated consumptive choices.[25] Barnard draws a parallel between this strategy and the labeling of cigarettes with information about the risks of smoking. The industry has resisted attempts at mandatory labeling and instead some companies have agreed to voluntary label programs. Currently, Certified Humane Raised and Handled is the largest animal welfare certification program.[26] These programs, however, frame the practices in a positive way, using terms such as "humane" and "free range," and the standards being followed to gain this and other forms of certification are frequently difficult to decipher. Labeling with more factual information that consumers could use in making purchasing decisions, or even including photographs of the production and processing of animals, would certainly be more informative, but would be difficult to get the industry to commit to.[27] Given the lobbying power of the industry, the implementation of mandatory labeling seems to be a distant dream.

If the strategies for improving animal welfare discussed above—educating consumers and using market demand, implementing new governmental policies, or using science and technology to leverage improvements in animal welfare—were successful in improving animal welfare domestically, they would necessarily increase costs in the short term at least, which would be passed on to consumers. The degree to which costs would increase varies by product and welfare modification. Matheny and Leahy provide the following estimates of the percentage cost increase of welfare improvements over standard practices: group housing sows, 0–3 percent; group housing calves, 1–2 percent; slow growth chickens, 5 percent; free-range turkeys, 30 percent; free-range hogs, 8–47 percent; furnished cages for laying chickens, 8–28 percent; barn environment for laying chickens, 8–24 percent; and free-range environment for laying chickens, 26–59 percent.

Some welfare improvements can increase productivity, but these increases would likely not outweigh the increased costs. Therefore, the only incentive producers will have to make animal welfare improvements is if demand is high enough to compensate for the increased costs and/or if they can pass along these costs, which consumers need to be willing to pay. That being said, there are two contextual factors that should be kept in mind. First, food costs per capita in the United States are currently lower than in almost every other country; however, this will likely be of little consolation to those who are currently struggling to purchase food for themselves and their families.[28] Second, while the cost of eating more humanely raised animals is certain to be higher, reducing consumption of meat products altogether and substituting with alternatives can in fact be cheaper.

The literature does indicate that the majority of consumers are in fact willing to pay marginally more for food that has some kind of animal welfare assurance. However, these willingness-to-pay claims do not necessarily translate into consistent purchasing behavior. This may be due to difficulties accessing such products, labeling that is insufficient for making an educated decision, or a belief that others (such as the state and industry) are responsible for ensuring animal welfare.[29]

Regardless of the reasons for the discrepancy between reported willingness to pay and actual consumptive behavior, the end problem is the same: If consumers are not willing to purchase products with animal welfare improvements then they will not be produced. Additionally, if improvements are implemented domestically, consumers might turn to cheaper imported animal products. This could result in a net aggregate decrease in animal welfare globally because the cheaper meat products would likely be imported from countries with much worse production and processing conditions.[30]

An alternative to this strategy of trying to convince producers to improve animal welfare is to instead pressure retailers, such as fast food restaurants and grocery stores, to modify their purchasing behavior. This strategy avoids some of the difficulties of consumers trying to decipher how companies are producing and processing in order to make educated purchasing decisions, as well as the problematic situation where imports from other countries with weaker welfare regulations gain a greater share of the market. Whereas "trade agreements can force nations to allow cheaper, non animal-friendly imports . . . they cannot force supermarkets or restaurants to sell them. Indeed, retailers may be more effective than regulators in affecting animal welfare."[31] Focusing efforts on retailers has resulted in significant gains to date. For instance, now McDonald's restaurants in the United Kingdom, Germany, Austria, and Switzerland use only free-range eggs, which has resulted in significant welfare improvements in the egg industry in those countries.[32]

The way it works is that the retailers set the standards that their producers have to meet. Then private or public organizations certify the claims made by producers. The reason it is effective is that retailers have significant power in today's market to demand changes because they control most of the access to the end consumer. For instance, McDonald's is the largest user of beef globally and is the first or second largest user of chicken.[33] Securing a contract with McDonald's is big business. Just as there has been growing consolidation in the animal production and processing industries, there has been significant consolidation in retailer companies in recent years. Producers are therefore vying for the large and lucrative contracts with these retailers, which have a lot of power in changing the rules of production and processing.[34]

Large retailers demanding better welfare standards among producers is a significant departure from previous practices, as retailers are at least partially responsible—if not primarily responsible, as argued by Bjerklie—for exerting pressure on producers and processors for reduced costs, which promoted increased industrialization in the first place. Retailers have also been relatively insulated from the repercussions of increasing industrialization; for instance, they have thus far avoided legal impacts of food contamination outbreaks at supplier facilities because they are not legally required to monitor what goes on in the production or processing arms of the process.[35] They have also largely escaped critical scrutiny related to the environmental impacts of meat production and the human health consequences (although there have been limited litigative attempts to hold them responsible). Heretofore, their status as "middlemen" has protected them, to a large degree. The power they have wielded, however, may now be usefully leveraged to improve animal welfare.

Targeting fast food chains has increasingly been used as a strategy by animal advocacy organizations in recent years. These organizations have used various means, including promoting consumer boycotts, influencing company shareholders, and targeting individual executives. Some direct action organizations have also used techniques such as vandalism. The targeting of McDonald's by animal advocates resulted in a notorious lawsuit in the United Kingdom in the 1990s, dubbed the "McLibel" lawsuit. McDonald's sued the activists, asserting that they were defaming the company in their campaign. The judge in the case ruled that some of the claims made by the activists in the written materials they distributed, such as the claim that McDonald's is responsible for world hunger, were libelous. Importantly, however, he also ruled that their claims that the company was implicated in the animal cruelty involved in the production and processing of the animals used in their products were not libelous because they were supported by fact. The judgement is important because

> in finding that McDonald's—the world's leading fast food provider—was "culpably responsible" for acts of animal cruelty, an impartial court of law established a clear, unambiguous legal connection between fast food providers and the methods used to produce animal-derived food for human consumption.[36]

In the wake of this ruling, and in response to further pressure from animal advocacy organizations, in recent years the major fast food companies in the United States, such as McDonald's, KFC, Burger King, and Wendy's, have implemented supplier requirements and created advisory boards to further develop and monitor production and processing requirements.[37]

Many commentators are optimistic about the potential of this strategy. Matheny and Leahy describe its efficacy as follows: "The visibility and name recognition of retailers make them sensitive targets of animal-welfare campaigns. As retailers compete with each other over public perception, successfully negotiating welfare gain with a major retailer can lead to a 'race to the top,' and to a push for harmonizing regulation so that costs are shared."[38] Although corporate executives are responsible for maximizing profits for shareholders, which made contracting with industrial producers and processors valuable in the past, if consumers become increasingly vocal about animal welfare, and vote with their dollars, acting progressively vis-à-vis animal welfare may become a new imperative among retailers because it is good for business in the long term.[39] Adams suggests that proactive retailers should use animal welfare to their competitive advantage and provide consumers with information about their welfare policies. They have yet to do so on a large scale, although there are some exceptions, such as the Chipotle chain of restaurants, which may indicate that the topic of animal welfare remains rather taboo among those who profit from the use of animals. Research is needed to quantify the effects the modest changes made thus far by retailers have actually had.[40]

SUSTAINABLE ANIMAL AGRICULTURE

The focus of the chapter up to this point has been on potential methods of improving animal welfare. Although some of the strategies for making industrial animal agriculture more sustainable may have positive effects on animal welfare, it is not always the case and it is not their specific purpose. Therefore, sustainability strategies are distinguished from methods targeted at animal welfare and discussed separately here.

As described throughout the previous chapters, industrial animal agriculture is unsustainable; this is little debated, particularly once the increasing consumption levels globally are taken into consideration. In short, "the quest for greater yields has landed farmers on a technologic treadmill of increasing inputs and decreasing profit margins."[41] These inputs include massive amounts of natural resources, particularly fossil fuels, water, and soil. The industry consumes these resources at an unrenewable/unsustainable rate. Past civilizations have actually collapsed because their farming utilized too many natural resources.[42] I would suggest that the industry is also increasingly socially unsustainable: too many negative impacts are being generated, and the costs of these impacts are not being borne by the industry and are instead being shouldered by others, including employees, consumers, neighbors, and broader communities.

Sustainable measures have been proposed to mitigate the treadmill industrial animal agriculture finds itself on. These include smaller farms that use fewer inputs from off the farm (e.g., pesticides), maintain biotic diversity, are more labor intensive, and use renewable forms of energy. These strategies run counter to many capitalist ideals, such as the equation of progress with industrialization and the focus on immediate profit maximization. Instead of focusing on the short-term goal of profit maximization, sustainable agriculture takes the long-term viability of production into consideration and consequently does not outstrip the resources available. Even though it takes a longer-term approach to profit-making, profitability is nonetheless one of the goals of sustainable agriculture, along with environmental health and social and economic equity.[43]

Farms can improve their sustainability by utilizing several methods different than those used by industrial agriculture. One such method is to have animals graze outside instead of being kept in confined, and often indoor, environments consuming industrially produced grains. Grazing is rotated so that animals are not contributing to soil erosion by overgrazing some areas. This method saves on feed costs, disposes of manure in a more distributive and safe way, and in doing so makes the soil more fertile.[44]

While this type of farming sounds good in theory, many are quick to question whether it is economically viable and able to compete with industrial agriculture. Direct comparisons are difficult to make, but the literature that is available indicates that sustainable agriculture can be competitive and that the demand for organic and sustainable food has increased dramatically in recent years: demand in the United States quadrupled between the years of 1997 and 2008 alone.[45] Community-supported agriculture (CSA), a model where customers buy a share of a farm and in exchange receive a portion of the farmed products, has proven to be a particularly useful model for sustainable farming because it provides farmers with capital up front, as well as a guaranteed market for their products.[46]

In addition to profitability, several other advantages are associated with this type of agriculture. First of all, this small, decentralized model puts individual farmers instead of corporations in control, thus mitigating the negative impacts of the corporatization of meat production and processing discussed in previous chapters. Second, this decentralized system of production and processing can improve food safety in the aggregate because if contamination does occur it is not amplified through large-scale cross-contamination.[47] Third, the environmental impacts of these decentralized farms are reduced. They utilize fewer renewable and nonrenewable resources and distribute the by-products (e.g., manure) in less concentrated ways.[48] Finally, animals benefit from a free-range grazing system more than intensive concentration.

Two qualifications regarding the welfare of animals in these systems are necessary: (1) organic standards alone do not necessarily mean that the animals are treated much differently than they are in industrial animal agriculture, and (2) even in instances where they are treated better, they are still eventually slaughtered, which necessarily involves harm.[49]

Despite the potential benefits, there are some reasons to be skeptical about the overall transformative potential of sustainable animal agriculture. Given the increasing levels of meat consumption and growing human population globally, meeting the demand for meat through sustainable meat production is not possible.[50] Even sustainable animal agriculture cannot escape the fact that meat production is inefficient: producing meat requires more calories in the form of inputs than it can deliver in an end product. Given this reality, some argue that the only solution to the massive negative impacts of industrial animal agriculture is a reduction in, or outright elimination of, meat consumption. Geographer Tony Weis is one of the advocates of less meat production and consumption. He writes, "Agricultural systems must be vastly more labor-intensive and biodiverse, and *geared towards much less meat production*, as finite biophysical overrides diminish and in order to reduce the untenable environmental and atmospheric burden of industrial methods."[51] Similarly, Ibrahim argues that because meat production and consumption are so inefficient, true sustainability requires a move away from animal consumption entirely. That strategy is discussed at the end of the chapter. Next we examine recommendations for changing regulation of the industry to mitigate impacts.

GOVERNMENTAL POLICIES

Some observers have suggested that voluntary national standards and labelling that reflects adherence to these standards would significantly improve practices in the industry and mitigate the negative impacts.[52] Although labels that are clear and explain the production and processing methods in a way that consumers can understand would certainly be an improvement, such labeling has not been implemented and has been resisted virulently by the industry. Against this backdrop, others have suggested that the government needs to play a more proactive role in regulating the industry and interceding in other ways to mitigate the negative impacts. Instead of exploring the gamut of suggestions for government intervention into industrial animal agriculture, I focus here on some of the more well-developed recommendations aimed at redressing particularly pernicious and significant problems within the industry, including introducing new legislation, better enforcing existing laws, modifying subsidy programs, and increasing targeted funding for research.

Introducing New Legislation

One suggestion that has garnered significant academic and public support is for the government to develop regulations that would require standards for the humane treatment of animals on farms. Farm animals are currently excluded from the Animal Welfare Act. Much of the general public does not realize that this is the case. Once advised of this legislative gap, the ratio of those who find this unacceptable to those who find it acceptable is more than three to one.[53] The industry would no doubt rebut that they have internal standards in place, that the slaughter of animals is regulated, and that there are humane production method certification programs consumers can purchase from if they so choose. Proponents of legislative measures in contrast argue that welfare laws that apply to the production of all livestock animals, and not just their deaths, would provide something that industry standards and voluntary certification programs simply cannot: transparency and consistency across the board.[54]

There have been proposals to enact laws regulating the welfare of animals in production at both the federal and state level. Mallon proposes the implementation of a Humane Standards of Living Act at the federal level, which would complement the current Humane Methods of Slaughter Act and provide welfare protections for animals while they are alive, not just when they are being killed. Mallon proposes that under this act, animals should be able to engage in natural behaviors, such as being outdoors and exercising, and that extremely restrictive cages (e.g., battery cages and gestation crates) be abolished. Others have recommended revising the Humane Methods of Slaughter Act to improve animal welfare in the processing arm of the industry. The recommendations include adding birds and fish in order to provide them with at least those protections and amending the act to specify maximum line speeds (i.e., the speed at which the disassembly line in slaughterhouses moves).[55]

A key advantage of creating new or revising current federal laws is that the treatment of farm animals would be consistent across the states, and the industry therefore would not be able to escape the regulations by moving from one state to another. There are also some notable disadvantages. The enforcement of the current Humane Methods of Slaughter Act is notoriously limited, and the points of enforcement are certainly fewer than the number of facilities producing and slaughtering animals. Second, as with the regulation of any industry, increased regulation could lead to an adversarial relationship between the industry and the regulators, which can make regulation more difficult in the end. Third, federal legislation can be rigid and difficult to change as the industry changes. Finally, there is no consensus in the animal sciences as to the minimum requirement of animal welfare and how we can assess it, which makes agreeing upon legislative standards more difficult.[56]

An alternative to creating or modifying federal law is to extend state anticruelty criminal laws to include livestock animals.[57] This would enable charging producers and processors criminally if their actions fall within the definition of cruel or abusive behavior. Presumably, this would remove the protection afforded by generally accepted practices within the industry. Extending state laws in this way would have the advantage of sending the message that as a society we are concerned about the abuse of not only companion animals, but also livestock animals. Vining argues that criminal laws would be particularly helpful because the perpetrators of the abuse in this context are corporations. He asks,

> "Should laws criminalizing animal abuse apply to animals raised for food?" The answer is yes, and yes especially because farm animals are generally now under the control of business corporations. State and federal criminal law have proved critical in modifying corporate policy and practice in other areas, a current example being worker safety. Criminal liability today would include criminal liability of the corporate entity itself, and would thus also introduce the most effective regulation of individual handling of farm animals—regulation by the corporate, which has methods and resources public agencies cannot match.[58]

He further suggests that such laws not be used against individual workers in the industry and instead target those with the power to make decisions about the treatment of the animals.

This strategy, however, also has associated limitations. First of all, state-by-state implementation of these laws would not be simultaneous, and producers could simply move to states where the anticruelty laws have not been extended in these ways.[59] Second, with these criminal laws intent must be proven. This may be particularly difficult to do within industrial animal agriculture because the perpetrators of the systemic cruelties are companies, which complicates the attribution of intent.[60] Third, there is a high burden of proof in criminal law (the charge must be proven beyond a reasonable doubt). Therefore, a significant amount of evidence will be necessary, and as has been pointed out in the context of corporate crime more generally, in order to gather the necessary evidence, some harms must be allowed to persist.[61] Fourth, prosecutors need to be willing to bring such cases forward for prosecution; thus, a broader cultural shift toward seeing livestock animals as worthy of protection as companion animals would be necessary.[62] Finally, the issue of enforcement cannot be overlooked. Enforcement of anticruelty laws in the context of abuses of beloved companion animals is notably weak; it is unclear how the enforcement of these extended laws would be any better.

Others have proposed a legal change that would have more of a sweeping impact on the industry than the previous two legislative approaches: changing the legal property status of animals.[63] Animals are able to be commodified and used by industrial animal agriculture to generate profit in the manner currently undertaken because they are legally considered property. Cassuto is among a growing number of scholars challenging this property status. His point of contention is that the divide between humans and animals enshrined in western law, and the designation of the latter as property, derives from the religious belief that a "morally relevant divide" exists between humans and animals. He argues that the blind acceptance and institutionalization of this religious precept requires critical legal examination. Space constraints prevent an examination here of the alternatives to property status for animals that legal scholars have proposed; suffice it to say that the alternatives proposed would significantly impact industrial animal agriculture and many other industries.[64]

In addition to proposed policy changes to improve animal welfare in industrial animal agriculture, there have also been many calls for governmental policy intervention to mitigate the pollution emitted by the industry. These have included general calls for getting tougher on pollution by the industry. For instance, Horrigan and colleagues recommend the implementation of stricter pollution policies and suggest that "without such policies, the products of factory farms will continue to be artificially cheap, in that prices will not reflect their impact on the environment, human health, animal welfare, or the economic and social stability of rural communities."[65]

More specifically, manure has become an area of focus. Manure poses one of the greatest environmental and health risks created by CAFOs, and a number of suggestions for improving its storage and disposal have been proposed. Some of the more proactive proposals include that the use of solid manure storage tanks be required instead of the lined lagoon style currently widely employed; manure go through waste treatment in order to mitigate contamination; all states require the licensing of manure handlers; approval of manure management plans be made mandatory and fines imposed if the plan is violated; ventilation systems in CAFOs be improved so that they filter out the air-borne organisms; and spreading manure on frozen ground be banned because it will not be absorbed and instead will simply run off.[66] Of course, these changes would increase the costs of production and are therefore not supported by the industry.

More reactive strategies for redressing the problems posed by manure include requiring producers to maintain insurance policies, put money into escrow accounts, or post bonds to cover the costs of environmental remediation if there is a problem.[67] Although these strategies are reactive, they may perform the indirect preventative function of making producers more cautious about their methods. Although not as threatening to the industry as preventative measures, such as those detailed above, they are also unlikely to be supported by the industry due to the potential increased costs.

The CAFO system has also been a focal point of the critical attention paid to industrialized animal agriculture. As the centralized point of production, it is the place from which many of the negative impacts emanate, particularly related to animal welfare, environmental contamination, and worker health and safety. Therefore, a number of suggestions have been made for changing the ways in which CAFOs are allowed to do business.

Due to the evidence of harms produced by existing CAFOs and the difficulty of undoing what has already been done, restricting the future construction of CAFOs has been suggested. The Workgroup on Community and Socioeconomic Issues recommends requiring public consultations before CAFO permits are issued. It has also been suggested that new CAFO developments be required to obtain a positive environmental impact statement.[68]

Other groups, such as the American Public Health Association, go further than recommending public consultations and environmental assessments and propose outright moratoriums on the construction of more CAFOs nationwide.[69] Academics have also voiced support for CAFO construction moratoriums. Such moratoriums are already in place in some states. For instance, North Carolina enacted a moratorium on pig CAFO construction; however, by the time they did so the pigs in the state outnumbered the human citizens. States face an uphill battle in implementing such moratoriums due to the lobbying power of the industry. Further, state-by-state implementation of moratoriums may simply shift the problem, as producers will simply move production to those states still welcoming CAFO construction. Even if moratoriums on future CAFO construction were implemented in all states, producers could simply relocate production operations to other countries. Smithfield already has operations in Poland and Romania, where regulations are currently even weaker than those in the United States.[70]

Although the proposed legislative actions discussed have the potential to mitigate the impacts of the industry, these strategies also have limitations worth addressing. First of all, the implementation of new policies—whether it be to improve animal welfare, reduce pollution by CAFOs, or for any other purpose—could be more difficult for smaller operations to comply with and cost them more relative to their production capacity than larger operations.[71] Second, policy change is extremely time consuming, and in the case of animal agriculture, progressive changes may be particularly difficult because of the close ties between the industry, politicians, and regulators.[72] Finally, as alluded to earlier, even if legislative changes were successfully made in an attempt to reduce the negative impacts of the industry, due to free trade, doing so could have the undesired consequence of worsening the impacts of the industry globally because producers and processors could simply relocate to countries with weaker laws and/or consumers could purchase imported animal products from countries with weaker laws.[73] Restrictions on imports or the application of tariffs could ameliorate this possibility, but they would likely be constructed as unfair barriers to trade by other countries and undoubtedly challenged by the World Trade Organization.[74]

Due to the difficulties associated with getting the government to make legislative changes, some animal advocacy organizations have turned to having referenda on specific industrial animal agriculture issues. This strategy has been successful in some states. For instance, sow gestation crates were banned in Florida in 2002, effective in 2008, and foie gras sale was banned in California in 2004, effective in 2012.[75] An infographic developed by the Animal Legal Defense Fund and aimed at educating the public about the production of foie gras and convincing voters to support the ban is captured in figure 28. Animal advocacy organizations, such as the Animal Legal Defense Fund, celebrated their success after voters supported their initiative and the ban was implemented (see figure 29).

This strategy, however, cannot escape some of the other limitations of legislative changes articulated above. Changes voted on by the people can be circumvented by consumers importing products from other states, which could also favor larger over smaller producers. Bans on the use or sale of products (such as foie gras) in a state could be more effective than production

Figure 28. Educating and mobilizing on the issue of foie gras production. (Source: ©Ian Elwood, Animal Legal Defense Fund. Used with permission.)

Figure 29. The Animal Legal Defense Fund celebrates the success of their campaign to ban foie gras in California. (Source: ©Ian Elwood, Animal Legal Defense Fund. Used with permission.)

bans, but these bans are also vulnerable to challenges on the grounds of free trade. The ability to replicate these referenda in other states is also uncertain, particularly because organizing for them is so resource intensive. The referendum in Florida alone cost animal advocacy groups more than $1 million. In spite of these limitations, these successful initiatives do send positive messages regarding the public's growing concern about farm animal welfare and how their food is produced more generally.[76]

Improve Enforcement of Existing Laws

Given the difficulties associated with developing new legislation to redress some of the problems inherent in industrial animal agriculture, perhaps a more preferable strategy would be to at least begin by better enforcing the laws that are currently on the books. As discussed in previous chapters, the regulations that do exist have been weakly enforced, and the industry has largely been left to regulate itself. Better enforcement of extant laws would likely serve a deterrent function. It would also assist in better assessing the various risks posed by the industry and would also even the playing field among those operations that are complying and incurring the costs to do so and those that save money through noncompliance.[77]

The specific recommendations made in the literature vary from identifying specific laws related to animal agriculture in need of better enforcement, improving whistle-blower provisions, making changes to regulatory jurisdiction, and better enforcing laws indirectly related to animal agriculture. Of the laws governing the industry, the Humane Methods of Slaughter Act stands out as being in need of better enforcement. One critic of current enforcement levels of the act remarks, "although many Americans believe that this 1958 law stops meat producers from inflicting unrelenting cruelty towards livestock, in actuality this law is not enforced at all and may as well not even be in existence."[78] Enforcement of this and other laws related to the industry could be made easier to some degree if changes were made to make it easier for whistle-blowers to come forward, such as making it possible for CAFO whistle-blowers to report anonymously to protect them from being penalized, which has happened in the past. Instead actions have been taken to make whistle-blowing in the industry more difficult, including proposals to make whistle-blowers pay for the inspection costs of their claims and even criminalizing whistle-blowing via "ag-gag" laws. Proponents of these measures imply that the risks posed by industrial animal agriculture are less serious than those created by other industries.[79] Such measures would not only negatively impact individual whistle-blowers, they would significantly reduce the amount of information the public receives about what goes on in the industry.

Making changes to the regulatory jurisdiction over the industry might also prove useful, as it is currently somewhat complicated. The USDA has the authority to monitor animal welfare in slaughterhouses and to create animal welfare standards within the Animal Welfare Act; however, the act explicitly excludes farm animals. Therefore, the USDA does not have a direct way to regulate the production of animals. The FDA, however, could be better situated to oversee at least some aspects of animal production because they have the authority to regulate the contents of livestock feed and livestock drugs and to control and prevent the spread of communicable diseases. If the FDA were to intervene in these issues in a meaningful way vis-à-vis livestock production, practices would invariably need to change.[80]

Not only does the USDA not have the authority to address the animal welfare of livestock while they are in the production sector of the industry, there have been questions raised about the agency's ability to regulate what they do have authority over, such as the Humane Methods of Slaughter Act. It has been suggested that the USDA is unable to properly regulate the animal agriculture industry due to two conflicts of interest: their mandate also includes promoting the industry that they are regulating, and the industry makes impressive political contributions, which has coincided with a significant number of appointees to the agency who are linked to the industry in various ways, particularly through past employment. In response to the current regulatory challenges, it has been suggested that some type of independent agency be created or that neutral third parties oversee the agencies that regulate the food supply.[81]

In addition to improving enforcement of current laws that most directly apply to animal agriculture, the enforcement of other laws indirectly related to the industry could prove useful. For instance, false-advertising laws could be used to dispute animal welfare claims, and antitrust laws could be invoked to challenge the high degree of consolidation in the industry, which has led to further industrialization. Invoking these indirectly related laws, as well as the laws directly governing the industry, and changing and/or improving their enforcement will require significant political will, much more than has been demonstrated to date.[82]

Amending Current Subsidy Policies

Not only does the current regulatory environment permit the harmful aspects of industrial animal agriculture to continue, the U.S. government also heavily subsidizes the industry.[83] As discussed in previous chapters, these subsidies take the form of providing price supports for animal products, subsidizing the production of animal feed, contributing to infrastructure, and subsidizing land use. Subsidizing grazing land for cattle ranchers alone costs taxpayers approximately $500 million a year. These subsidies make it possible for the industry to charge consumers less for their final product, but they also facilitate the overuse of resources and the other negative industry impacts.[84]

Proponents of farm subsidies argue that they help small family farmers who might otherwise be driven out of the market. What is less commonly acknowledged is that the current system of subsidies actually favors large, corporate, industrial farms instead of small farmers.[85] The International Institute for Sustainable Development analyzed these subsidies and concludes that "almost 30% of subsidies go to the top 2% and over four-fifths to the top 30%. Ironically, if the United States government were to shift its target from the top 30% to the bottom 70% of farmers, it could save at least $8 billion a year while supplying a competitive boost to lower-income farms."[86]

If subsidies are going to continue to be applied to the industry, they should be applied to ensure that small farming operation profits exceed their costs of production, and as an incentive for the industry to improve animal welfare, its environmental sustainability, and workplace conditions.[87] This strategy, however, faces the same challenges as the implementation of new legislation and better enforcement of existing regulations: there must be a political will to do so, and the lobbying power of the industry has certainly dampened political will in the past.

Increase and Strategically Target Government Funding

More research on industrial animal agriculture, its impacts, and potential alternatives would also prove beneficial. Schmidt suggests that the current dearth of research on the impacts of industrial animal agriculture is at least partially attributable to the difficult position in which many researchers find themselves: if they report negative findings about the industry they can reduce their chances of securing further research funding. Thus, according to Robert Martin, a senior officer with the Pew Environmental Group, "even the best scientists seem loathe to say anything against the industry. With the decline in public research funding, it's industrial animal agriculture that pays for virtually all of the animal sciences research going on at land-grant universities today."[88] In addition to funding challenges, difficulties securing access to the industry can also limit research on industrial animal agriculture. Mench and colleagues identify a pressing need for animal welfare research on functioning commercial farms in particular; however, they cite difficulties gaining access and maintaining researcher independence as posing significant barriers.

Government funding for animal agriculture research is currently focused on finding technological fixes to industrial problems instead of addressing and redressing the root of the unsustainability of industrial animal agriculture.[89] Research on the potential alternatives to industrial animal agriculture is also needed. Such research does not have the corporate backing that research on industrial animal agriculture does. This problem could be ameliorated if government funding was directed to research that supports sustainable agriculture instead of research that further benefits industrial animal agriculture.[90] In a period of fiscal tightening, funding research into alternatives to industrial animal agriculture is even more unlikely.

International Standards

Even if government subsidies were restructured, research into alternatives to industrial animal agriculture were better funded, and government-imposed or publicly voted-on changes in regulations of the types discussed were implemented, U.S. meat production and processing companies could relocate (to other countries) and in many cases consumers can import (cheaper) products from other jurisdictions with weaker regulations. Regional or international standards would mitigate this looming problem. The way it would work in theory is that countries would consult and agree to implement a set of minimum standards for the industry in their country.

There are two examples related to animal welfare that illustrate what this type of approach looks like. One example was mentioned earlier: the Protocol on Animal Welfare (part of the Treaty of Amsterdam), signed by European states in 1997. Among other things, it recognizes animals as sentient beings. To date, however, this recognition has changed little in practice.[91] The second example I wish to highlight here is the international guidelines for animal welfare implemented in 2005 by the World Organisation for Animal Health (OIE), which represents 178 countries. The organization adopted welfare standards related to the transportation of livestock animals, slaughter, and killing livestock animals for the purpose of mitigating the spread of disease. The guidelines are rather conservative, but what is noteworthy is that the organization can be referred to in cases of World Trade Organization dispute settlements, meaning the recommendations could actually influence international trade.[92] A conference focused on supporting the implementation of the guidelines in member countries was held in 2012. How exactly these guidelines will be implemented and the impacts thereof will be known in the near future. It will no doubt set the tone for future international agreements around animal agriculture.

In addition to establishing regional or international guidelines or standards, another technique for improving industrial animal agriculture internationally, particularly as it emerges in developing countries, is to modify the requirements for the allocation of assistance for animal agriculture in developing countries provided by international grants and loans such that industrial developments are not favored.[93] Some progress is being made in this regard. In 2001, the World Bank made a commitment to avoid supporting industrial animal agriculture.[94] Research examining the implementation of this commitment and the impact to date in various national contexts would provide valuable information about this strategy for mitigating the impacts of industrial animal agriculture globally.

IMPROVING WORKER HEALTH AND SAFETY

Historically, unions within the industrial animal agriculture industry were able to secure some improvements to working conditions, such as worker access to bathrooms as needed.[95] The declining power of the unions in the second half of the twentieth century, however, has left workers more vulnerable. The industry has also been known to scapegoat workers for problems that arise in facilities.[96] In general, increased unionization in the industry has been identified as a way to protect workers from scapegoating and assist with identifying some of the underlying problems in the workplace that can compromise worker health and safety.

Along with this general call for increased unionization, other more specific strategies for improving worker health and safety (which could be facilitated by greater unionization) have been proposed. These suggestions include improving worker training, which would not only improve worker safety, but would also likely improve animal welfare and food safety; hiring people to work in the production of livestock animals based on their attitudes toward animals, because research has demonstrated that workers' attitudes toward animals are good predictors of job performance and treatment of animals; providing workers with influenza vaccinations to protect them against contracting influenza viruses and preventing the spread of viruses to other workers by encouraging workers to stay home when ill; and providing workers with respirators and on-site showers so that workers can reduce the amount of contaminants they come into contact with and expose others to.[97]

It is not just the physical well-being of workers that should be of concern; attention should also be paid to their emotional and mental well-being. Working inside CAFOs and witnessing daily the pain and suffering caused by the extreme confinement may have a negative impact on some people; research is needed to assess this possibility. The potential harmful consequences of slaughterhouse work on the mental well-being of workers have only recently been acknowledged.[98] In response to the emerging research, Dillard suggests that slaughterhouse work be designated an ultrahazardous activity in order to give workers increased access to compensation due to the potential psychological impacts.[99] Further research is needed in the production and processing sectors of the industry to determine the scope of the problems experienced by workers and to devise strategies to mitigate the impact on workers.

IMPROVING MEAT SAFETY

Some of the suggestions discussed above, such as improving worker training, could also serve to improve the safety of meat products; however, there are additional and more specific suggestions for improving food safety that deserve attention here. There are three strategies that would arguably have the most significant impacts. The first is to implement changes in the diets of livestock animals. This would include more stringent restrictions on the inclusion of animal by-products in animal feed, ideally implementing strict vegetarian diets for livestock animals, as well as improving enforcement of the regulations that are in place. Doing so would reduce the possibility of spreading zoonotic illnesses via food.[100] Additionally, requiring that cattle are grass-fed would reduce the risk of contamination by the dangerous variant of E. coli bacteria.[101]

The second strategy is to begin a serious evaluation of the current way that meat is processed, which includes participants from the corporate, governmental, academic, labor, and public sectors. As discussed in the previous chapters, due to concentration and industrialization in the industry, there are fewer slaughterhouses in the country than there used to be, and these slaughterhouses kill and process hundreds of animals an hour. Killing and processing this many animals per hour makes it more likely that the contaminated remains of one animal will spread to others. Slowing down the line speeds would likely not only reduce meat contamination, but would also improve animal welfare and worker health and safety.[102] Additionally, processed meat products can include the meat from many different animals. Therefore, the risk is even further distributed. The current method of controlling meat contamination, the hazard analysis and critical control points system, which was supposed to mitigate the problem, has had the perhaps unintended effect of exacerbating it by further increasing corporate concentration in the industry and amplifying contaminants.[103] Experience has demonstrated that technological and regulatory changes aimed at reducing contamination in the industry are unlikely to eliminate or even significantly reduce the problem; instead there is a need to critically examine the problems inherent in the structure of the industry.

The third and more commonly proposed way to improve the meat supply is to reduce antibiotic use in livestock production. Although antibiotics are not necessarily considered food contaminants, they can have long-term human health consequences, such as antibiotic resistance. The European Union has already banned nontherapeutic use of antibiotics in livestock production, and the World Health Organization (WHO) has implored other countries to ban this practice as well. Gilchrist and colleagues suggest that, instead of waiting for countries to act on this, the industry should be proactive in reducing their use of antibiotics, and that companies already doing so should be identified and used as examples for others in the industry to emulate.[104]

These suggestions for improving the safety of animal-derived products are viewed by the industry with skepticism and met with resistance, particularly because there will no doubt be costs involved. The position of the industry in this regard would not surprise many people; however, more people would likely be surprised to learn that some animal advocates are also unsupportive of such changes. Whereas animal welfarists are supportive of positive changes in the industry, such as slowing down line speeds, at least some animal rightists and liberationists would argue that such changes merely perpetuate a system that kills billions of animals. In short, tinkering with production and processing does not keep animals from being killed, only attacking the consumption part of the equation can do that.

CHALLENGING THE CONCEPTUALIZATION OF ANIMALS AS FOOD: VEGETARIANISM AND VEGANISM

The suggestions discussed up to this point have, for the most part, been relatively specific and focused on the production and processing sectors of the industry. The strategies discussed are also primarily aimed at working within the confines of the industry, although some of the proposals push the boundaries somewhat. The last two strategies I would like to discuss in this chapter are more radical in that they seek to destabilize the foundation of the industry by abstaining from consuming its products, and they challenge the culturally pervasive ideologies related to industrial animal agriculture and animals more generally.

After decades of the public paying very little attention to what they were being fed, recently significant numbers of people have begun to reflect upon what they are consuming and where it comes from. Books, films, magazines, and media reports about food are proliferating. Social movements have developed around issues such as organic food, humane production of animal products, community gardening, cooperative agriculture, and vegetarianism/veganism. What all of these movements share in common is that they are opening up space to critically reflect upon the way that industrialized agriculture produces food.[105]

As discussed earlier in the chapter, switching out the consumption of industrially produced meat for the consumption of more sustainably produced meat can mitigate some of the negative impacts of industrialized food production.[106] However, there is reason for skepticism about how "humane" more sustainable modes of producing meat can claim to be. It is impossible to eliminate animal suffering from the equation—at the very least, animals are still slaughtered.[107] Humane animal agriculture also leaves the property status of animals and speciesism unchallenged: animals are still treated as commodities, which leaves them vulnerable to harm.[108] Even if the negative impacts of industrial animal agriculture could be eliminated through sustainable/humane meat production, logistical problems abound. In particular, it seems unlikely that the high demand for meat could be met through this means. Additionally, the cost would be prohibitive for many. The skeptics, such as Ibrahim, therefore suggest there really are only two options: "because animal foods can only be made affordable for most consumers through factory farming, society is left with a dichotomous choice: either we stop purchasing and consuming animals, or they will continue to suffer in our factory farms."[109] In short, sustainable agriculture will be unable to produce meat products as cheaply or to sustain the pressure of the current level of demand.

In light of that fundamental problem, a growing number of people are concluding that the only way to eventually eliminate the animal welfare problems posed by animal agriculture, along with the other significant negative impacts of the industry, is to stop consuming animal products altogether. Philosopher Tom Regan articulates this perspective as follows:

> the way to end the injustice of slaughter is not through reform or regulation. Injustice reformed is and always will be justice delayed. There can be no end to the injustice of slaughter until there is an end to the business of slaughter. Because corporations that make fortunes off the death of hogs, chickens and other animals will never close their doors for reasons of conscience, the slaughterhouse walls must be dissembled, one brick at a time, by acts of conscience performed by one person at a time. The only way to stop the supply of meat is to eliminate the demand for meat.[110]

Thus, Regan would likely argue that the suggestions aimed at making modifications to the process of meat production and processing discussed throughout this chapter will not stop all of the negative impacts of the industry; the only way to do so is to target consumption and advocate for mass vegetarianism and veganism.

This strategy for mitigating the negative impacts of industrial animal agriculture is not without its own critics. Michael Pollan's view, outlined in his book *The Omnivore's Dilemma* and elsewhere, is that there is a fundamental tension between concern for animals and concern for the environment and that decisions made in favor of animals compromise the environment. He suggests that the adoption of vegetarian diets in some geographic regions, such as the Arctic, would result in environmental harm because vegetarian food would have to be shipped in. Stănescu counters that Pollan's argument overlooks the fact that the environmental costs of producing meat to meet current demand outweighs the costs of transporting vegetarian items to various geographic areas. He also argues that a decentralized system of meat production of the type Pollan suggests would further harm the environment because more land would be needed.[111]

The debate over the merit of refraining from eating meat is likely as old as the practice itself. Vegetarianism has a long history, which really cannot be done justice within the confines of the pages of this book, so a brief overview will have to suffice. Although as far back as Ancient Greece people were abstaining from eating meat, including Pythagoras and Plato, vegetarianism did not gain significant momentum until the nineteenth century.[112] At this time, people were drawn to it due to ethical concerns about consuming animals, as well as a belief that consuming animals polluted one's body.[113] By the turn of the twentieth century, the vegetarian population was still rather small and was mainly populated by groups as diverse as intellectuals, the working class, and nonconformist branches of Christianity.[114] Negative perceptions of vegetarians (e.g., as weak and eccentric) persisted through the first half of the twentieth century, which no doubt prevented many people from considering it as an option.[115]

The popularity of vegetarianism increased again beginning in the 1960s. It began to gain momentum at this time as part of the antiestablishment counterculture movement. It was further fueled through the 1970s and 1980s by scientific studies that began to document a connection between the consumption of animal products and several human health ailments. The growing number of vegetarians at this time created a niche market, which promoted the development of vegetarian products. Accessibility of these products, particularly in urban areas, consequently made it more convenient for others to become vegetarian. In the 1990s, crises in meat safety, such as the emergence of mad cow disease, influenced more people to give vegetarianism a try.[116]

Perceptions of vegetarians have certainly improved over time. They are no longer widely considered heretics, a charge levelled during the Inquisition by the Catholic church. Research indicates vegans and vegetarians now tend to be viewed as being moral and principled. Despite this significant improvement in reputation, some negative stereotypes of vegetarians persist today, such as the perception that they are weak (particularly male vegetarians/vegans).[117]

Determining the number of people who abstained from meat historically is fraught with difficulty. Even today, determining the proportion of the population following meat-free diets is difficult, particularly because people do not always behave consistently with what they report.[118] Additionally, people employ different definitions of vegetarianism; for instance, some people consider those who eat fish and/or eat "white meat" vegetarians.[119] There are some generally agreed upon definitions, although certainly not everyone is aware of them: vegetarians who eat eggs and dairy are referred to as lacto-ovo vegetarians; the term lacto-vegetarian refers to those who abstain from consuming eggs, but will consume dairy; and ovo-vegetarians will consume eggs but not dairy. Vegans are strict vegetarians who refrain from consuming all animal products or using them in other ways, including as clothing.

In spite of these difficulties, surveys provide us with rough estimates of the proportion of the population who abstain from eating meat. According to one estimate, approximately 3 percent of the U.S. adult population does not consume meat. Studies have documented a similar proportion of vegetarians in the United Kingdom, but higher proportions in other countries, such as Canada (8 percent), Germany (9 percent), Israel (8.5 percent), and India (40 percent). Other statistics indicate that the proportion of those who abstain from meat consumption in the United States is likely to increase: one quarter of the population in the United States reports that they are currently reducing their meat consumption, 10 percent of the population is following a more generally worded "vegetarian-inclined" diet, and 5 percent express that they are considering future vegetarianism.[120]

Studies have also examined why people decide to eschew meat products, and all animal-derived products in some cases, from their diets. It is important to keep in mind, though, that often people adopt vegetarianism or veganism for more than one reason, and the reason for adhering to this dietary change can fluctuate over time.[121] Recent studies indicate that ethical concerns are the most commonly reported motivator.[122] The ethical concerns center on the instrumentalization of other beings and violating their rights by consuming them. In the past few decades, the treatment of animals used as food has been identified as an issue that should be of specific concern to feminists.[123] Building on intersectional feminist thinking, which proposes that forms of oppression such as sexism, racism, classism, and heterosexism are interconnected and mutually reinforcing, some feminists (notably ecofeminists) argue that the oppression of women and animals is interconnected: patriarchy and speciesism reinforce each other. Therefore, they assert that consuming animals indirectly reinforces the patriarchal system, which among many of its characteristics values culture over nature. From this perspective, Lynda Birke argues, "politics that ignore other oppressions cannot be liberatory politics for anyone."[124]

Recent research indicates that health concerns are the second most commonly cited motivation for becoming vegetarian or vegan.[125] Interestingly, however, this motivation may have a gendered component. A study of undergraduate students found that male students are more likely to invoke health as a reason to eat meat instead of to abstain from it, indicating that at least for this age group the health factor may be more persuasive for women than men.[126]

Researchers have also studied the barriers that exist to becoming vegetarian or vegan. These include (in descending order of reported importance) enjoyment of meat, unwillingness to alter consumption habits, belief that humans are meant to eat meat, consumption of meat by family members, and inadequate knowledge about vegetarianism and veganism. Gender also plays a role in the perception of these barriers. Specifically, research indicates that men are more likely to believe that humans are meant to consume meat and see this as an insurmountable barrier, and women are more likely to invoke their family's consumption of meat as a reason for not adopting a meat-free diet.[127]

Gender is not only related to the reasons people become vegetarians or vegans and the reasons they do not, it is also reflected in the membership overall, with more women than men reporting to be vegetarian/vegan.[128] Kalof and colleagues argue based on their empirical findings that more women than men abstain from meat due to differing values and beliefs. Altruism in particular is important in mediating this relationship. Additionally, they find that white men stand out in being skeptical about the belief that vegetarian diets are beneficial.[129] Their finding that in the aggregate altruism plays an important role in the path toward vegetarianism is consistent with other research that finds that vegetarians score higher on measures of empathy toward other people than omnivores, and that individuals who are motivated to abstain from consuming meat for ethical reasons exhibit significantly higher use of areas of the brain associated with empathy when actually observing human and animal suffering.[130]

The ability to mobilize this empathy and altruism beyond individual consumptive behaviors toward agitation for broader social change has, however, been limited. In her book on vegetarianism, Maurer argues that, far from being simply a lifestyle, vegetarianism is a social movement, complete with organizations, its own movement literature, philosophical principles, products and services, and ideology. No doubt many individuals—particularly those concerned not only with what enters their body but also with the countless animals who are produced, processed, and consumed by others—would consider themselves part of this social movement. However, there is a great deal of vegetarianism/veganism that gets distilled down into individual consumerism.[131] For Gunderson, this is akin to the problem of green consumerism, where concern for the environment becomes individualized, people do not question their consumption, and instead purchase what is marketed by companies as "green." Those who are advocating for broader social change argue that this lifestyle or consumerist form of vegetarianism/veganism is problematic because those who insulate themselves from meat products can feel as though they are doing their part without actually challenging the root of the problem.[132]

While vegetarianism and veganism can be a powerful way of attending to the linkages between consumption and production and processing, when it is approached solely as a lifestyle issue, the reintegration of production, processing, and consumption is relegated to being an issue of individual choice: if framed as a lifestyle issue, individual choice becomes foregrounded.[133] Proponents of vegetarianism/veganism as part of a broader social movement, therefore, would disagree with the answer Michael Pollan provides in his book *The Omnivore's Dilemma* in response to the question of whether consuming the products of industrial animal agriculture is acceptable. He states, "In the end each of us has to decide for himself [*sic*] whether eating animals that [*sic*] have died in this manner is okay."[134] The challenge facing the current vegetarian/vegan movement is that it has to convince people that their individual dietary decisions can have a positive impact, while not focusing only on the individual benefits of meat-free diets, agitating for broader social and economic changes, and recruiting those into the movement who are going to work toward realizing those larger goals.[135]

This begs the following question: What strategies should the vegetarian/vegan movement employ to convince others to give up meat and become part of a social movement to challenge industrial animal agriculture? Some suggest that the message of the movement should be focused on the risks associated with consuming meat; however, individual risks should not be the only focus, for the reasons just discussed.[136] Others have suggested that movement messages should be tailored to appeal to women and men. In his research on justifications for meat consumption and masculinity, Rothgerber finds that men tend to use more direct ways of justifying meat consumption (e.g., employing health and religious reasons), whereas women tend to use more indirect methods, such as avoiding thinking about the fact that meat comes from animals and the ways that these animals were treated. He therefore suggests that both the direct and indirect ways of justifying meat consumption need to be addressed, and more generally, it would be worthwhile to educate men about the way that they have been socialized to connect meat consumption with masculinity. Therefore, a more nuanced approach would be more beneficial than attempting to simply educate the general public about what goes on within the walls of CAFOs and slaughterhouses.[137]

Another recommended strategy is to develop connections with other social movements around common interests. The vegetarian/vegan movement should reach out to interested parties across the realms of production, processing, and consumption, such as consumer groups, immigration advocates, labor unions, groups concerned with food issues more generally, the environmental movement, and the animal advocacy movement more generally.[138] All of these interests could powerfully coalesce to form a new social movement focused on challenging industrial animal agriculture.[139]

In addition to engaging in strategic alliances, the vegetarian/vegan movement and the anti–industrial animal agriculture movement more specifically would do well to focus their attention strategically on specific issues. Marcus proposes four areas where this anti–industrial animal agriculture movement should first strategically focus. The first two were recommended earlier in this chapter: modifying subsidies (he cites grain subsidies in particular) to the industry so that the price of meat comes closer to representing the costs of producing it; and critically examining the role of the USDA, with the specific goal of transferring responsibility for developing nutritional guidelines away from the USDA, to the National Institutes of Health. He also recommends reforming school lunch programs to make them more nutritious and more vegetarian-focused. This could simultaneously improve the well-being of children and reduce the short-term and long-term demand for industrial animal agriculture products. Related to this, he also suggests focusing attention on young people. These individuals might be more receptive to vegetarianism/veganism, and giving up meat at a young age saves more animals over a lifetime. He estimates that a college-aged individual giving up meat saves approximately two thousand animals. Convincing five hundred therefore saves a million animals.

I would add focusing on the contribution of the industry to global climate change to Marcus's list of strategic areas worth pursuing. In doing so, animal advocates can leverage the significant concern that already exists for mitigating global climate change as another reason to eschew animal products from one's diet and accentuate the fact that doing so will outweigh even the positive effects of changing one's mode of transportation. While such specific and targeted strategies can be helpful in focusing social movement energy, it is also useful to keep the larger, cultural context in mind.

CHANGING CULTURAL IDEOLOGY

Hardeman and Jochemsen's thorough analysis of the ideological elements of modern agriculture was detailed in a previous chapter. In short, as a result of their research they suggest that many ideological traits are evident in the industry, including the presence of an absolutized end (i.e., efficiency); reformulation of ideas, values, and norms to be consistent with the ideology; using the absolutized end to select the method (which has resulted in intensification and specialization); rationalization; radicalization; instrumentalization; attempts to solve problems utilizing the same methods that contributed to the original problems; and the use of ideological or religious terms to refer to industrialized agriculture. Their conclusion is that industrialized agriculture contains significant ideological elements; a conclusion that might be surprising to the casual observer.[140]

These ideological elements constrain the way that we see the system and our ability to develop and appreciate alternatives to it. Geographer Tony Weis uses the example of the way that productivity is measured in the industry to demonstrate the important yet shrouded role that ideology plays:

> Though it is not always understood in this way, the conceptualization of agricultural productivity—and with it, the efficiency gains of industrial agriculture—is an ideological exercise. . . . A very different conception of productivity would emerge, including a much greater range of measurable costs and "outputs," if agricultural systems were designed with goals such as minimizing GHG [greenhouse gas] emissions, soil erosion, toxicity. . . . The need to radically restructure agriculture is at the very core of any hope of making the "human relation to the earth" more ecologically sustainable.[141]

The difficulty is that various ideological elements, especially the ones Weis and Hardeman and Jochemsen identify, envelop and insulate the industry, making it more difficult to question the current system of industrial animal agriculture in the first place and to propose alternatives to it with the goal of "radically restructur[ing] agriculture," as Weis puts it.

For truly significant change to take place it is necessary to begin to challenge these ideological elements. One way to do so, which is a goal of this book, is to literally and symbolically bring the processes of production, processing, and consumption—and the consequences thereof—into closer contact. The current separation of these processes is neither accidental nor innocent: as Gouveia and Juska put it, it is "an artifact of power and socio-cultural as well as ideological construction."[142] Beginning to bridge the chasm between these processes will challenge the ideological elements detailed above that have served to protect the industry from widespread critical attention.

Further, bridging the divide between the processes of production, processing, consumption, and their impacts will also bring different constituencies and social movement organizations into closer contact around issues that are of common interest, but heretofore have largely not been viewed as such or have even been viewed antagonistically. In particular, there is quite a bit of common ground related to the impacts of industrial animal agriculture where environmental and animal advocacy groups could come together. Yet, to date, a degree of divisiveness on the issue remains. Empirical analyses of the outreach materials of environmental organizations have found that when these organizations do attend to the problems posed by industrialized animal agriculture they tend not to address the animal welfare issues. For instance, an author of a book on factory hog farms states in the prefatory material, "I've purposely avoided the debate over animal welfare because it would have detracted from the more important issues facing Nebraska."[143] We need a more integrated response that does not bracket out certain impacts and certain actors. When environmental groups do recommend altering meat consumption patterns, they suggest eating "sustainable meat" and rarely address the possibility of reducing or eliminating meat consumption. They mostly blame agricultural companies for the negative environmental impacts but say little about the role consumers play in creating demand for these products.[144]

Making the connections between the processes of production, processing, consumption, and their impacts will also require a politicization of the issue among the citizenry. There is growing awareness of some of the negative consequences that can befall consumers of meat, such as contracting a food-borne illness. However, the negative impacts of the industry are not only borne by people in their capacity as consumers. They are also borne by people in their capacity as workers and even by people in their capacity as citizens. Kendall Thu goes as far as to argue that industrialized animal agriculture is a threat to democracy because citizens do not have a say in the system, even though the negative impacts are affecting them. It is not just a consumer issue: it needs to be politicized as a social, species, and environmental justice issue impacting all citizens.[145]

Elevating these issues in our collective cultural consciousness will not be easy, particularly because, as Joy points out, our culture is resistant to confronting topics that elicit discomfort. Addressing industrial animal agriculture requires us to tackle the realities of how animals in the industry live and die, which is particularly difficult to do. She explains, "Bearing witness means choosing to suffer. Indeed, empathy is literally 'feeling with.' Choosing to suffer is particularly difficult in a culture that is addicted to comfort—a culture that teaches that pain should be avoided whenever possible and that ignorance is bliss."[146]

THE FUTURE . . .

I acknowledge that up to this point in this book, the picture painted has been fairly bleak. The students I teach in my classes frequently remark that the more they learn, particularly about the issues discussed in this book, the more difficult it is for them not to become pessimistic. Despite the size and scope of the problems discussed in this book, however, I think there is reason for optimism that the ills wrought by industrial animal agriculture will garner greater attention and positive changes will be made, perhaps out of absolute necessity. I envision a convergence of factors—technological changes, environmental limits, political alliances, changing public sentiments, and legal challenges—that have the potential to change the shape of the industry.

In the past, technological developments served the interests of the industry and furthered the industrialization process. Today, there is a technological development on the horizon that may actually work against the interests of the animal agriculture industry. So-called test-tube or in vitro meat is currently being developed. In the summer of 2013, the first hamburger made from muscle stem cells from a cow and grown in a lab was cooked and sampled. Those who tasted the burger indicated the taste was close to a regular burger, but not exactly there yet. The cost of developing this burger was an astounding $300,000 plus; however, further development and higher levels of production would certainly reduce the cost dramatically. Scientists are working on refining the product, and they estimate that it will be ready for commercial sale in ten to twenty years. This product could nearly eliminate the animal welfare and environmental costs of producing meat.[147] Of course, in order for these products to become widely available and challenge the dominance of industrially produced meat, production will have to be deemed profitable, and in order for that to be the case, individuals will have to demonstrate a willingness to consume meat that is manufactured in a lab instead of in CAFOs and slaughterhouses.[148] Educating consumers will play an important role in this regard. Having the topic of test-tube meat on the media's radar at the very least brings the issue of meat production and processing into the public's consciousness.

We can also reasonably expect that industrial animal agriculture will further be brought to the public's attention as the resources needed by the industry become increasingly scarce and the environmental consequences of the industry become more apparent. As demonstrated herein, there is an overwhelming amount of evidence that the industry is unsustainable. Even though the animals they use are "renewed," many of the resources it uses, particularly fossil fuels, are not renewable, while the manure generated by the industry and its effects live on and on. No doubt social movement organizations will have an important role to play in this consciousness raising.

Weis predicts that it is the scarcity of fossil fuels that will be the first thing to dramatically impact the industry, which relies on relatively cheap oil for activities such as transportation and the production of feed. As oil becomes increasingly scarce and expensive, food costs will increase, which Weis predicts will lead to increasing malnourishment and food riots. The world has already begun to witness this. But Weis does not think that all of the consequences will be negative. He anticipates that positive change may be born from this impending crisis, writing, "a period of acute instability looms, potentially for much worse but also potentially for better. As in all systemic crises, there is hope that seams for radical change can widen."[149]

Others are not as hopeful about potential radical change. In her book on animal capital and biopolitics, Shukin states that she is not optimistic that the internal contradictions of capitalism more generally will destabilize and ultimately destroy it. She argues that it is all too capable of surviving and is similarly pessimistic about the possibility of significant change in industrial animal agriculture. She acknowledges that there are many problems evident in the industry, but because "they are not yet a formulation of political struggle," she suggests there is little hope for meaningful change.[150]

While I appreciate Shukin's caution, I share in Weis's somewhat cautious optimism. We are witnessing the seeds of political agitation around the issue of industrial animal agriculture. For instance, a joint environmental social justice movement has emerged around the siting of CAFOs in communities.[151] As the community consequences of slaughterhouses receive increased attention, we might expect that a similar movement will develop around the siting of those facilities as well.[152] As the negative impacts of industrial animal agriculture and their connections to other social and environmental causes become more apparent, the coalition building between movements around this issue will hopefully be fostered. The negative impacts affect several different constituencies—environmental advocates, animal advocates, labor unions, small farmers, community health advocates, and immigrant worker advocates, to name a few—that combined have significant resources and reach. Again, Weis's words are instructive: "far from fading away into modernity, struggles over agriculture are bound to intensify as the promise of cheap industrial food breaks down, and might well come to occupy a pivotal place in broader anti-systemic movements at the precipice of mitigation and adaptation or climate change disaster."[153] The industrial animal agriculture system has the potential to bring social movements together on issues of global scope, such as global climate change, and of local concern, such as the opening of new CAFOs in communities.

In addition to technological changes, environmental limits, and political alliances that have the potential to put pressure on the industry, increasing agitation might also emerge from evolving sentiments among the general public. Recall historian Richard Bulliet's premise that we have moved from what he refers to as the domestic era, where people have frequent contact with various types of animals, to the postdomestic (beginning in the 1970s). In the postdomestic era, a growing distance (physically and psychologically) is created between people and the animals they use, yet paradoxically, people form increasingly close relationships with their pets.

The sociohistorical forces Bulliet describes have resulted in a growing tension between the love the majority of the population feels for their pets and the indifference toward livestock animals. Bulliet writes that postdomestic citizens "experience feelings of guilt, shame, and disgust when they think (as seldom as possible) about the industrial processes by which domestic animals are rendered into products."[154] The indifference toward livestock animals is facilitated by a culture that makes it possible, as Bulliet states, to think about it "as seldom as possible." We might expect that the more people are confronted by the realities of industrial animal agriculture—whether their attention is drawn to it through undercover videos of animal cruelty in CAFOs and slaughterhouses, the possibility of in vitro meat, crises caused by product recalls due to meat contamination, or the spread of zoonotic illnesses that originate in CAFOs—the greater the tension will become. The industry may be able to get undercover videos criminalized or frame in vitro meat so negatively that the consumer base will not purchase it, but without addressing the structural problems inherent in the industry, it will continue to engender crises. The connections between production, processing, consumption, and impacts will be exposed. As Horowitz remarks, "The incomplete victory over nature by our contemporary provisioning system guarantees a steady stream of controversies over meat."[155] Further, the tension exposed by these controversies and crises will only be exacerbated as the number of animals in industrial animal agriculture increases and their quality of life simultaneously declines even further.[156]

My argument that, despite the power and best efforts of the industry, the connections between the supply, consumption, and impacts of animal-derived products will increasingly be exposed and significant changes made as a result may be considered by some to be overly optimistic. Yet there is evidence that the industry is concerned about the transformative potential of the factors identified here. In his textbook on animal welfare and meat production, Gregory poses the following question: "Should the meat and livestock industry be worried about animal rights?"[157] By way of response, he points to the fact that the European Union recognized animals as sentient beings in the 1990s and warns that this could have legal implications. As I write this, the World Trade Organization has just upheld a ban on the importation of Canadian seal products into the European Union. Although they found that the ban does violate trade agreements, they ruled that this violation is acceptable because the Canadian seal hunt offends the "public morals" in the European Union.[158] The potential exists to use this reasoning in future decisions about the trade in animal agricultural products. Attitudes toward animals in the postdomestic era are not only changing, they are beginning to figure into legal decisions.[159]

With all of this being said, it is still impossible to predict exactly what the future of meat production, processing, and consumption will hold; however, we can be fairly certain that "as with other domains of risk management in contemporary society, the place of meat and its future are fundamentally unstable."[160] As evidenced throughout this book, the production, processing, and consumption of meat has changed significantly over human history, and at a dramatic pace in the most recent past. In spite of these significant changes it is tempting to normalize the current methods of producing, processing, and consuming animals as food because it is what we are now most familiar with—for some, it is indeed the only method they know. The result is a cultural perception of the current practices of production, processing, and consumption as stable, normal, and inevitable. The conceptual and literal distance that has been created between production, processing, and consumption is also normalized and taken for granted. This distancing has made concealment possible and has obscured not only the public's view of the industry, but also sectors of the industry from each other.[161] As a result, there is a need to educate ourselves about the ways animals used as food are produced and processed, how this has changed, the consequences thereof, and importantly *why* the current methods of production, processing, and consumption are used today. This understanding destabilizes a normative view of current methods and makes it possible to envision alternatives.

This destabilization, which this book indicates is likely to increase, makes it more likely that industrial animal agriculture will assume a more prominent place on the public's radar and the political agenda, even if out of necessity. Although this instability may be increasing, it is certainly not new. In his book *Man and the Natural World*, historian Keith Thomas foregrounds a particularly pernicious source of instability stemming from conflict between our affective relations with animals and our use of them as food. He writes,

> A mixture of compromise and concealment has so far prevented this conflict from having to be fully resolved. But the issue cannot be completely evaded and it can be relied upon to recur. It is one of the contradictions upon which modern civilization may be said to rest. About its ultimate consequences we can only speculate.[162]

Cracks in the facade of concealment—combined with the impacts of technological changes, environmental limits, shifting political alliances, evolving public sentiments, and mounting legal challenges—will only augment this instability.

Notes

INTRODUCTION

1. Lorber, *Paradoxes of Gender*, 13.
2. See Berger and Luckmann, *Social Construction of Reality*.
3. C. Adams, "Ecofeminism and the Eating of Animals" (1996), 135.
4. Magdoff, Foster, and Buttel, *Hungry for Profit*; Bulliet, *Hunters, Herders, and Hamburgers*; Gunderson, "From Cattle to Capital."
5. Singer and Mason, *The Way We Eat*.
6. Lappé, "Diet for a Hot Planet."
7. Gregory, *Animal Welfare and Meat Production*.
8. Gunderson, "From Cattle to Capital," 260.
9. Pollan, *Omnivore's Dilemma*, 305.
10. *National Hog Farmer*, "Consumer Survey Outlines Pork Perception Issues."
11. Mayfield et al., "Consumption of Welfare-Friendly Food Products."
12. Pollan, *Omnivore's Dilemma*; Purdy, "Open the Slaughterhouses."
13. Vialles, *Animal to Edible*, 5.
14. Weis, "Accelerating Biophysical Contradictions of Industrial Capitalist Agriculture," 317–18.
15. Singer and Mason, *The Way We Eat*.
16. Scully, "Fear Factories."
17. R. Adams, "Fast Food and Animal Rights"; Gunderson, "From Cattle to Capital"; Nibert, *Animal Rights/Human Rights*; Rifkin, *Beyond Beef*.
18. Mench et al., "Welfare of Animals in Concentrated Animal Feeding Operations," 7.
19. Pollan, *Omnivore's Dilemma*; Foer, *Eating Animals*.
20. Mench et al. "Welfare of Animals in Concentrated Animal Feeding Operations."
21. Marcus, *Meat Market*.
22. Sapontzis, *Food for Thought*.
23. Dunayer, *Animal Equality*.
24. Clay, *World Agriculture and the Environment*.
25. Mazoyer and Roudart, *History of World Agriculture*; Sabin, *Agriculture*; Woods and Woods, *Ancient Agriculture*.
26. Woods and Woods, *Ancient Agriculture*, 6.
27. FAO, "State of Food and Agriculture."
28. Singer and Mason, *The Way We Eat*.

CHAPTER 1. PREHISTORY THROUGH THE
COLONIZATION OF NORTH AMERICA

1. Mithen, "Hunter-Gatherer Prehistory of Human-Animal Interactions."
2. Mithen, "Hunter-Gatherer Prehistory of Human-Animal Interactions"; Cartmill, *A View to a Death in the Morning.*
3. Mithen, "Hunter-Gatherer Prehistory of Human-Animal Interactions," 197.
4. Kalof, *Looking at Animals in Human History*; Mithen, "Hunter-Gatherer Prehistory of Human-Animal Interactions"; Cartmill, *A View to a Death in the Morning.*
5. Cartmill, *A View to a Death in the Morning.*
6. Longo and Malone, "Meat, Medicine, and Materialism," 115.
7. Noske, *Humans and Other Animals*, 6.
8. Noske, *Humans and Other Animals.*
9. Wilkie, "Sentient Commodities and Productive Paradoxes"; Mithen, "Hunter-Gatherer Prehistory of Human-Animal Interactions."
10. Kalof, *Looking at Animals in Human History*; Mithen, "Hunter-Gatherer Prehistory of Human-Animal Interactions."
11. Mithen, "Hunter-Gatherer Prehistory of Human-Animal Interactions"; Kalof, *Looking at Animals in Human History.* The assertion that horses were domesticated at this early time in human history is controversial.
12. Wilkie, "Sentient Commodities and Productive Paradoxes."
13. Clutton-Brock, *Animals as Domesticates*; Wilkie, "Sentient Commodities and Productive Paradoxes."
14. Wilkie, "Sentient Commodities and Productive Paradoxes."
15. Clutton-Brock, *Animals as Domesticates*, 24.
16. Clutton-Brock, *Animals as Domesticates.*
17. Kalof, *Looking at Animals in Human History.*
18. Nibert, *Animal Oppression and Human Violence*, 12.
19. Noske, *Humans and Other Animals*; Price and Gebauer, "New Perspectives on the Transition to Agriculture"; Clutton-Brock, *Animals as Domesticates.*
20. Kalof, *Looking at Animals in Human History*; Rifkin, *Beyond Beef.*
21. Rifkin, *Beyond Beef*, 19.
22. Cattle continue to play an important role in rituals in some cultures today. For instance, the Zulus in southern Africa use cow feces and hair in rituals with human newborns; they are washed with cow feces and receive a necklace made from cow hair (Rifkin, *Beyond Beef*). Cattle also continue to have a special status in Hinduism.
23. Rifkin, *Beyond Beef*, 23.
24. Rifkin, *Beyond Beef.*
25. Kalof, *Looking at Animals in Human History*; Nibert, *Animal Oppression and Human Violence.*
26. Rifkin, *Beyond Beef*, 28.
27. Price and Gebauer, "New Perspectives on the Transition to Agriculture"; Nibert, *Animal Oppression and Human Violence.*
28. Wood, *Origin of Capitalism.*
29. The role of agriculture, of course, was not the only reason that England transitioned to capitalism. A number of factors played a role. For instance, it was a relatively integrated and politically centralized country (Wood, *Origin of Capitalism*).

30. See also Headlee, *Political Economy of the Family Farm* and Nibert, *Animal Oppression and Human Violence.*

31. Longo and Malone, "Meat, Medicine, and Materialism"; Nibert, *Animal Oppression and Human Violence.*

32. Noske, *Humans and Other Animals*; Longo and Malone, "Meat, Medicine, and Materialism," 118.

33. For instance Chambers and Mingay, *The Agricultural Revolution*; Thompson, "Second Agricultural Revolution."

34. Wood, *Origin of Capitalism*; Noske, *Humans and Other Animals.*

35. Thompson, "Second Agricultural Revolution."

36. Wood, *Origin of Capitalism.*

37. Kalof, *Looking at Animals in Human History.*

38. Wallerstein, *Modern World-System I.*

39. Thompson, "Second Agricultural Revolution," 63.

40. Thompson, "Second Agricultural Revolution."

41. Thompson, "Second Agricultural Revolution."

42. Wood, *Origin of Capitalism*, 97.

43. Clutton-Brock, *Animals as Domesticates.*

44. Rifkin, *Beyond Beef.*

45. Clutton-Brock, *Animals as Domesticates.*

46. Rifkin, *Beyond Beef.*

47. Clutton-Brock, *Animals as Domesticates.*

48. Anderson quoted in Wilkie, *Livestock/Deadstock*, 27.

49. Mithen, "Hunter-Gatherer Prehistory of Human-Animal Interactions," 201.

CHAPTER 2. THE INDUSTRIALIZATION OF LIVESTOCK PRODUCTION

1. Thompson, "Second Agricultural Revolution."

2. Hardeman and Jochemsen, "Are There Ideological Aspects to the Modernization of Agriculture?"

3. McNeil, *An Encyclopedia of the History of Technology.*

4. McNeil, *An Encyclopedia of the History of Technology.*

5. Clutton-Brock, *Animals as Domesticates.*

6. Hardeman and Jochemsen, "Are There Ideological Aspects to the Modernization of Agriculture?"

7. Franklin, *Animals and Modern Cultures.*

8. Skaggs, *Prime Cut.*

9. Rifkin, *Beyond Beef.*

10. Rifkin, *Beyond Beef.*

11. Striffler, *Chicken.*

12. Horowitz, *Putting Meat on the American Table*; Striffler, *Chicken.*

13. Skaggs, *Prime Cut.*

14. Winders and Nibert, "Consuming the Surplus."

15. Winders and Nibert, "Consuming the Surplus."

16. D. Fitzgerald, *Every Farm a Factory*; Winders and Nibert, "Consuming the Surplus."

17. Winders and Nibert, "Consuming the Surplus."

18. Winders and Nibert, "Consuming the Surplus."

19. Cited in Winders and Nibert, "Consuming the Surplus," 80.
20. Winders and Nibert, "Consuming the Surplus"; Horowitz, *Putting Meat on the American Table.*
21. The regulatory definition of a CAFO depends upon the type of animal being produced, the number of animals in the facility, and whether the animals and/or their manure come into contact with surface water. To give the reader a sense of how large these facilities can be, the number of animals in large CAFOs are included here. Large CAFOs contain 1,000 or more cattle; 2,500 or more pigs weighing more than 55 pounds or 10,000 or more pigs weighing less than 55 pounds; 55,000 or more turkeys; 30,000 or more laying hens or broilers with a liquid manure handling system or 82,000 hens or more with alternative manure handling system (EPA, "Regulatory Definitions of Large CAFOs, Medium CAFOs, and Small CAFOs").
22. Ibrahim, "A Return to Descartes," 101.
23. Horowitz, *Putting Meat on the American Table.*
24. Wilkie, "Sentient Commodities and Productive Paradoxes"; Ibrahim, "A Return to Descartes."
25. Horowitz, *Putting Meat on the American Table.*
26. Horowitz, *Putting Meat on the American Table*; Starmer, Witteman, and Wise, "Feeding the Factory Farm."
27. Starmer, Witteman, and Wise, "Feeding the Factory Farm"; Striffler, *Chicken.*
28. Hahn Niman, *Righteous Porkchop*; Goldschmidt, *As You Sow.*
29. Horowitz, *Putting Meat on the American Table.*
30. Ibrahim, "A Return to Descartes," 93.
31. Starmer, Witteman, and Wise, "Feeding the Factory Farm"; Horowitz, *Putting Meat on the American Table*; Ibrahim, "A Return to Descartes."
32. Horowitz, *Putting Meat on the American Table*; Starmer, Witteman, and Wise, "Feeding the Factory Farm."
33. Starmer, Witteman, and Wise, "Feeding the Factory Farm," 14.
34. Starmer, Witteman, and Wise, "Feeding the Factory Farm," 13.
35. Starmer, Witteman, and Wise, "Feeding the Factory Farm," 15.
36. Skaggs, *Prime Cut.*
37. For a detailed discussion, see Ibrahim, "A Return to Descartes."
38. U.S. Poultry and Egg Association, "Economic Data."
39. Horowitz, *Putting Meat on the American Table*; Ibrahim, "A Return to Descartes."
40. Horowitz, *Putting Meat on the American Table.*
41. Johnsen, *Raising a Stink.*
42. Burmeister, "Lagoons, Litter and the Law."
43. McBride and Key, "U.S. Hog Production from 1992 to 2009."
44. Horowitz, *Putting Meat on the American Table.*
45. Horowitz, *Putting Meat on the American Table.*
46. Dunayer, *Animal Equality.*
47. Gunderson, "From Cattle to Capital."
48. Ibrahim, "A Return to Descartes."
49. Cited in Marcus, *Meat Market*, 8.
50. Gunderson, "Metabolic Rifts of Livestock Agribusiness," 405.
51. Starmer, Witteman, and Wise, "Feeding the Factory Farm."
52. Starmer, Witteman, and Wise, "Feeding the Factory Farm," 4.
53. Starmer, Witteman, and Wise, "Feeding the Factory Farm," 7.
54. Starmer, Witteman, and Wise, "Feeding the Factory Farm."

55. Starmer, Witteman, and Wise, "Feeding the Factory Farm."
56. Starmer, Witteman, and Wise, "Feeding the Factory Farm."
57. Environmental Working Group, "Farm Subsidies."
58. Hennessy, "Slaughterhouse Rules."
59. Ibrahim, "A Return to Descartes."
60. Hennessy, "Slaughterhouse Rules."
61. Twine, *Animals as Biotechnology*.
62. Clutton-Brock, *Animals as Domesticates*.
63. Twine, *Animals as Biotechnology*.
64. Twine, *Animals as Biotechnology*; Abbott, "Pig Geneticists Go the Whole Hog."
65. Twine, *Animals as Biotechnology*, 63.
66. Twine, *Animals as Biotechnology*.
67. University of Guelph, "Enviropig"; Clark, "Ecological Biopower, Environmental Violence against Animals, and the 'Greening' of the Factory Farm."
68. Perkel, "University of Guelph 'Enviropigs' Put Down."
69. Twine, *Animals as Biotechnology*.
70. Twine, "Searching for the 'Win-Win'?"
71. Gunderson, "From Cattle to Capital," 260.
72. Gunderson, "From Cattle to Capital"; Wood, *Origin of Capitalism*.
73. Clutton-Brock, *Animals as Domesticates*.
74. Gunderson, "From Cattle to Capital," 267.
75. Gunderson, "From Cattle to Capital."
76. Schnaiberg, *The Environment*.
77. Novek, "Intensive Hog Farming in Manitoba."
78. Burmeister, "Lagoons, Litter and the Law"; Zuzworsky, "From the Marketplace to the Dinner Plate."
79. Bjerklie, "Size Matters"; Mench et al., "Welfare of Animals in Concentrated Animal Feeding Operations"; Ransom, "Rise of Agricultural Animal Welfare Standards."
80. Hardeman and Jochemsen, "Are There Ideological Aspects to the Modernization of Agriculture?"
81. Hardeman and Jochemsen, "Are There Ideological Aspects to the Modernization of Agriculture?"
82. Hardeman and Jochemsen, "Are There Ideological Aspects to the Modernization of Agriculture?," 669.
83. Weis, "Accelerating Biophysical Contradictions of Industrial Capitalist Agriculture."
84. Noske, *Humans and Other Animals*.
85. Twine, "Revealing the 'Animal Industrial Complex,'" 23.
86. Twine usefully points to the parallels between the animal-industrial complex and the prison-industrial complex, entertainment-industrial complex, and pharmaceutical-industrial complex. What they share in common is that each is able to turn misery into profit, and "although all of these have been naturalized by a familiar ideology that would root them in an assumed 'human nature' (as has capitalism itself) they in fact reveal major tensions between corporate agency and democratic oversight" ("Revealing the 'Animal Industrial Complex,'" 19).
87. Noske, *Humans and Other Animals*, 35.
88. Weis, "Accelerating Biophysical Contradictions of Industrial Capitalist Agriculture," 320.
89. Babad, "Apples Shares Tumble."
90. Nierenberg, "Factory Farming in the Developing World."
91. Matheny and Leahy, "Farm-Animal Welfare, Legislation, and Trade."
92. Twine, *Animals as Biotechnology*.

93. Burmeister, "Lagoons, Litter and the Law."
94. Twine, *Animals as Biotechnology*, 129.
95. Nierenberg, "Factory Farming in the Developing World."
96. Matheny and Leahy, "Farm-Animal Welfare, Legislation, and Trade."
97. Wilkie, "Sentient Commodities and Productive Paradoxes."
98. Twine, "Revealing the 'Animal Industrial Complex.'"
99. Matheny and Leahy, "Farm-Animal Welfare, Legislation, and Trade."
100. Nierenberg, "Factory Farming in the Developing World."
101. Nierenberg, "Factory Farming in the Developing World."
102. Nibert, *Animal Rights/Human Rights*; Winders and Nibert, "Consuming the Surplus."
103. Twine, *Animals as Biotechnology*.
104. Nierenberg, "Factory Farming in the Developing World."
105. Ponette-González and Fry, "Pig Pandemic."
106. Nierenberg, "Factory Farming in the Developing World."
107. D. Fitzgerald, *Every Farm a Factory*, 189.
108. Matheny and Leahy, "Farm-Animal Welfare, Legislation, and Trade."
109. Nierenberg, "Factory Farming in the Developing World"; Zande, "Raising a Stink."
110. Burmeister, "Lagoons, Litter and the Law"; Zande, "Raising a Stink."
111. Zande, "Raising a Stink."
112. Zande, "Raising a Stink."
113. Burmeister, "Lagoons, Litter and the Law."
114. Burmeister, "Lagoons, Litter and the Law."
115. Johnsen, *Raising a Stink*.
116. Hahn Niman, *Righteous Porkchop*.
117. Zande, "Raising a Stink," 25.
118. Stith, Warrick, and Sill, "Boss Hog."
119. Hahn Niman, *Righteous Porkchop*.
120. Stith, Warrick, and Sill, "Boss Hog."
121. Stith, Warrick, and Sill, "Boss Hog."
122. Kennedy, "Foreword."
123. Kennedy, "Foreword."
124. *Boston Globe*, "Animal Cruelty."
125. *New York Times*, "Eating with Our Eyes Closed."
126. Atwood, *The Edible Woman*.
127. Foer, *Eating Animals*.
128. Foer, *Eating Animals*, 130.
129. Such treatment of animals inspired the aptly named book *Animal Machines*, by Ruth Harrison, published in 1964. The book is credited as being one of the first to address the conditions inside industrial animal agriculture.
130. Dunayer, *Animal Equality*.
131. Mason and Finelli, "Brave New Farm?"; Ibrahim, "A Return to Descartes."
132. Mason and Finelli, "Brave New Farm?"
133. Mason and Finelli, "Brave New Farm?"
134. Mason and Finelli, "Brave New Farm?"
135. Eisnitz, *Slaughterhouse*.
136. Horowitz, *Putting Meat on the American Table*.

137. Mason and Finelli, "Brave New Farm?"; Foer, *Eating Animals*.
138. Dunayer, *Animal Equality*.
139. Mason and Finelli, "Brave New Farm?"
140. Mason and Finelli, "Brave New Farm?," 109.
141. Mason and Finelli, "Brave New Farm?"
142. Dunayer, *Animal Equality*, 131.
143. For thorough discussions of the lives of farmed fish, see Mason and Finelli, "Brave New Farm?" and Foer, *Eating Animals*.
144. Foer, *Eating Animals*, 131.
145. Ibrahim, "A Return to Descartes."

CHAPTER 3. THE INDUSTRIALIZATION OF SLAUGHTER AND PROCESSING

1. Vialles, *Animal to Edible*.
2. Otter, "Civilizing Slaughter."
3. Dunayer, *Animal Equality*.
4. Fitzgerald, Kalof, and Dietz, "Slaughterhouses and Increased Crime Rates"; A. Fitzgerald, "A Social History of the Slaughterhouse."
5. York, "Humanity and Inhumanity," 263.
6. Ibrahim, "A Return to Descartes," 103; Marcus, *Meat Market*.
7. One rather concrete linkage between the realms of animal production and processing is the transportation of animals from where they are raised (primarily CAFOs) to the slaughterhouse. This chapter addresses the ways in which technological developments in transportation contributed to changes in the slaughtering industry; however, there is not enough space to detail how animals are currently transported to slaughter and the problems these processes entail. For a more detailed discussion of those issues, see Marcus, *Meat Market*.
8. Horowitz, *Putting Meat on the American Table*.
9. Kay, "Butchers, Tanners, and Tallow Chandlers."
10. MacLachlan, "Humanitarian Reform, Slaughter Technology, and Butcher Resistance," and Brantz, "Animal Bodies, Human Health, and the Reform of Slaughterhouses."
11. Brantz, "Animal Bodies, Human Health, and the Reform of Slaughterhouses."
12. Skaggs, *Prime Cut*.
13. Skaggs, *Prime Cut*; Azzam, "Competition in the US Meatpacking Industry"; Patterson, *Eternal Treblinka*.
14. Horowitz, *Putting Meat on the American Table*.
15. Otter, "Civilizing Slaughter"; Brantz, "Animal Bodies, Human Health, and the Reform of Slaughterhouses."
16. MacLachlan, "Humanitarian Reform, Slaughter Technology, and Butcher Resistance."
17. Brantz, "Animal Bodies, Human Health, and the Reform of Slaughterhouses"; Vialles, *Animal to Edible*.
18. Vialles, *Animal to Edible*, 72.
19. Elias, *The Civilizing Process*.
20. Thomas, *Man and the Natural World*.
21. A. Fitzgerald, "A Social History of the Slaughterhouse."

22. Thomas, *Man and the Natural World*.
23. Philo, "Animals, Geography, and the City," 63, 65.
24. Kalof, *Looking at Animals in Human History*.
25. Thomas, *Man and the Natural World*, 300–301.
26. Thomas, *Man and the Natural World*.
27. Thomas, *Man and the Natural World*; Young Lee, "Siting the Slaughterhouse."
28. Vialles, *Animal to Edible*, 15 (emphasis mine).
29. A. Fitzgerald, "A Social History of the Slaughterhouse."
30. Hughes, "Good People and Dirty Work"; Vialles, *Animal to Edible*.
31. Kay, "Butchers, Tanners, and Tallow Chandlers"; Horowitz, *Putting Meat on the American Table*.
32. Horowitz, *Putting Meat on the American Table*.
33. Horowitz, *Putting Meat on the American Table*, 26.
34. Young Lee, "Siting the Slaughterhouse"; Twine, *Animals as Biotechnology*. See also Serpell, *In the Company of Animals* and Smith, "The 'Ethical' Space of the Abattoir."
35. A. Fitzgerald, "A Social History of the Slaughterhouse."
36. Cronon, *Nature's Metropolis*.
37. Vialles, *Animal to Edible*, 22.
38. Rifkin, *Beyond Beef*.
39. Horowitz, *Putting Meat on the American Table*.
40. Rifkin, *Beyond Beef*.
41. Horowitz, *Putting Meat on the American Table*.
42. Rifkin, *Beyond Beef*; Skaggs, *Prime Cut*.
43. For a detailed history of how meatpacking in Chicago forever changed meat processing and the city, see Cronon (*Nature's Metropolis*), Horowitz (*Putting Meat on the American Table*), Jablonsky (*Pride in the Jungle*), Patterson (*Eternal Treblinka*), and Skaggs (*Prime Cut*).
44. Rifkin, *Beyond Beef*.
45. Jablonsky, *Pride in the Jungle*.
46. Horowitz, *Putting Meat on the American Table*.
47. Rifkin, *Beyond Beef*.
48. Skaggs, *Prime Cut*; Wilkie, "Sentient Commodities and Productive Paradoxes."
49. Wilkie, *Livestock/Deadstock*, 31.
50. Wilkie, "Sentient Commodities and Productive Paradoxes."
51. Horowitz, *Putting Meat on the American Table*.
52. Skaggs, *Prime Cut*.
53. Cronon, *Nature's Metropolis*, 207.
54. Horowitz, *Putting Meat on the American Table*, 29.
55. Horowitz, *Putting Meat on the American Table*.
56. Horowitz, *Putting Meat on the American Table*.
57. Horowitz, *Putting Meat on the American Table*, 108.
58. Rifkin, *Beyond Beef*.
59. Wilkie, *Livestock/Deadstock*, 31.
60. Horowitz, *Putting Meat on the American Table*; Shukin, *Animal Capital*; Rifkin, *Beyond Beef*.
61. Rifkin, *Beyond Beef*; Shukin, *Animal Capital*.
62. Pacyga, "Chicago: Slaughterhouse to the World"; Patterson, *Eternal Treblinka*; Rifkin, *Beyond Beef*; Stull and Broadway, "Effects of Restructuring on Beefpacking in Kansas."
63. Rifkin, *Beyond Beef*.

64. Rifkin, *Beyond Beef*.
65. Wilkie, "Sentient Commodities and Productive Paradoxes."
66. Vialles, *Animal to Edible*; Hughes, "Good People and Dirty Work."
67. Gouveia and Juska, "Taming Nature, Taming Workers," 370.
68. Shukin, *Animal Capital*.
69. Quoted in Horowitz, *Putting Meat on the American Table*, 50.
70. Cronon, *Nature's Metropolis*, 208.
71. Sinclair, *The Jungle*; Skaggs, *Prime Cut*. For further details on the slaughtering process and conditions inside slaughterhouses at this time, see Horowitz, *Putting Meat on the American Table*.
72. Rifkin, *Beyond Beef*.
73. Skaggs, *Prime Cut*; Rifkin, *Beyond Beef*.
74. Rifkin, *Beyond Beef*; Skaggs, *Prime Cut*.
75. Skaggs, *Prime Cut*; Rifkin, *Beyond Beef*.
76. Skaggs, *Prime Cut*.
77. Skaggs, *Prime Cut*, 88–89.
78. Skaggs, *Prime Cut*; Horowitz, *Putting Meat on the American Table*.
79. Skaggs, *Prime Cut*.
80. Skaggs, *Prime Cut*; Horowitz, *Putting Meat on the American Table*.
81. Rifkin, *Beyond Beef*.
82. Skaggs, *Prime Cut*.
83. Rifkin, *Beyond Beef*.
84. Rifkin, *Beyond Beef*.
85. Skaggs, *Prime Cut*; National Labor Relations Board, "National Labor Relations Act."
86. Horowitz, *Putting Meat on the American Table*; Halpern, *Down on the Killing Floor*; Horowitz, "'Where Men Will Not Work.'"
87. Brueggemann and Brown, "Decline of Industrial Unionism in the Meatpacking Industry."
88. IBP was purchased by Occidental Petroleum in 1981 (Skaggs, *Prime Cut*; and Rifkin, *Beyond Beef*) and then by Tyson Foods in 2001 (Stull and Broadway, *Slaughterhouse Blues*.)
89. Skaggs, *Prime Cut*.
90. Rifkin, *Beyond Beef*; Stull and Broadway, *Slaughterhouse Blues*; Gouveia and Juska, "Taming Nature, Taming Workers."
91. Gouveia and Juska, "Taming Nature, Taming Workers"; Rifkin, *Beyond Beef*.
92. Azzam, "Competition in the US Meatpacking Industry"; Brueggemann and Brown, "Decline of Industrial Unionism in the Meatpacking Industry"; Stull and Broadway, *Slaughterhouse Blues*; Stull and Broadway, "Effects of Restructuring on Beefpacking in Kansas."
93. Rifkin, *Beyond Beef*.
94. Stull and Broadway, "Effects of Restructuring on Beefpacking in Kansas."
95. Right-to-work laws prohibit requiring employees to join or pay dues to a union and make it possible for union members to resign from the union at any time. Employees who are not members of the union, however, still enjoy the benefits of the collective agreement. These laws therefore essentially remove the incentive to pay dues and be a member of a union, thereby weakening unionization. Rifkin, *Beyond Beef*; Azzam, "Competition in the US Meatpacking Industry."
96. Skaggs, *Prime Cut*.
97. Skaggs, *Prime Cut*; Rifkin, *Beyond Beef*.
98. Broadway and Stull, "Meat Processing and Garden City, KS"; Dickes and Dickes, "Oligopolists Then and Now."

99. Broadway, "Where's the Beef?"
100. Pachirat, *Every Twelve Seconds*, 22.
101. Brueggemann and Brown, "Decline of Industrial Unionism in the Meatpacking Industry."
102. Stull and Broadway, "Effects of Restructuring on Beefpacking in Kansas."
103. Skaggs, *Prime Cut*; Rifkin, *Beyond Beef*.
104. Rifkin, *Beyond Beef*.
105. Horowitz, *Putting Meat on the American Table*.
106. Brueggemann and Brown, "Decline of Industrial Unionism in the Meatpacking Industry."
107. Rifkin, *Beyond Beef*.
108. Benson, "Effects of Packinghouse Work on Southeast Asian Refugee Families."
109. U.S. Department of Labor, Bureau of Labor Statistics, "Labor Force Statistics from the Current Population Survey."
110. U.S. Government Accountability Office, "Workplace Safety and Health."
111. Gouveia and Juska, "Taming Nature, Taming Workers," 379.
112. Gouveia and Juska, "Taming Nature, Taming Workers."
113. Gouveia and Juska, "Taming Nature, Taming Workers."
114. Gouveia and Juska, "Taming Nature, Taming Workers."
115. Gouveia and Juska, "Taming Nature, Taming Workers," 377.
116. Gouveia and Juska, "Taming Nature, Taming Workers."
117. Brueggemann and Brown, "Decline of Industrial Unionism in the Meatpacking Industry."
118. Horowitz, *Putting Meat on the American Table*.
119. Horowitz, *Putting Meat on the American Table*.
120. Ibrahim, "A Return to Descartes," 96.
121. Skaggs, *Prime Cut*.
122. Ibrahim, "A Return to Descartes."
123. U.S. Poultry and Egg Association, "Economic Data."
124. Horowitz, *Putting Meat on the American Table*.
125. Cited in Skaggs, *Prime Cut*, 218.
126. Horowitz, *Putting Meat on the American Table*, 149.
127. Horowitz, *Putting Meat on the American Table*.
128. Horowitz, *Putting Meat on the American Table*, 150.
129. Brueggemann and Brown, "Decline of Industrial Unionism in the Meatpacking Industry."
130. Marcus, *Meat Market*.
131. Marcus, *Meat Market*, 7.
132. Wilkie, "Sentient Commodities and Productive Paradoxes."
133. Wilkie, *Livestock/Deadstock*, 173.
134. Gouveia and Juska, "Taming Nature, Taming Workers."
135. Vialles, *Animal to Edible*.
136. Vialles, *Animal to Edible*; Gouveia and Juska, "Taming Nature, Taming Workers."
137. See Foer, *Eating Animals*; Pollan, *Omnivore's Dilemma*.
138. Pachirat, *Every Twelve Seconds*, 14.
139. Skaggs, *Prime Cut*; Rifkin, *Beyond Beef*.
140. Rifkin, *Beyond Beef*.
141. Marcus, *Meat Market*.
142. USDA, "Humane Methods of Slaughter Act."
143. Gouveia and Juska, "Taming Nature, Taming Workers."

144. Gouveia and Juska, "Taming Nature, Taming Workers," 382.

145. Gouveia and Juska, "Taming Nature, Taming Workers."

146. Gouveia and Juska, "Taming Nature, Taming Workers," 383.

147. Gouveia and Juska, "Taming Nature, Taming Workers."

148. Quoted in Rifkin, *Beyond Beef*, 139.

149. Marcus, *Meat Market*.

150. Shukin, *Animal Capital*, 25.

151. Gregory, *Animal Welfare and Meat Production*; Pachirat, *Every Twelve Seconds*.

152. Gregory, *Animal Welfare and Meat Production*; Marcus, *Meat Market*; Horowitz, *Putting Meat on the American Table*.

153. Marcus, *Meat Market*.

154. Gregory, *Animal Welfare and Meat Production*.

155. Pachirat, *Every Twelve Seconds*.

156. Horowitz, *Putting Meat on the American Table*; Marcus, *Meat Market*; Ibrahim, "A Return to Descartes."

157. Pearce, Sheridan, and Bolton, "Distribution of Airborne Microorganisms in Commercial Pork Slaughter Processes."

158. Marcus, *Meat Market*.

159. See for example Schlosser, *Fast Food Nation*.

160. Warrick, "'They Die Piece by Piece,'" B8.

161. Warrick, "'They Die Piece by Piece,'" B8.

162. Warrick, "'They Die Piece by Piece.'"

163. Warrick, "'They Die Piece by Piece.'"

164. Pachirat, *Every Twelve Seconds*.

165. Eisnitz, *Slaughterhouse*, 88 (emphasis mine).

166. Vialles, *Animal to Edible*.

CHAPTER 4. CONSUMING ANIMALS AS FOOD

1. C. Adams, "Ecofeminism and the Eating of Animals" (1996), 130.

2. Fitzgerald and Taylor, "Cultural Hegemony of Meat and the Animal Industrial Complex."

3. Foer, *Eating Animals*; Pollan, *Omnivore's Dilemma*.

4. Gouveia and Juska, "Taming Nature, Taming Workers," 371.

5. Gouveia and Juska, "Taming Nature, Taming Workers." See also Fitzgerald and Baralt, "Media Constructions of Responsibility for the Production and Mitigation of Environmental Harms."

6. Dauvergne, *Shadows of Consumption*, 165.

7. Franklin, *Animals and Modern Cultures*, 145.

8. Franklin, *Animals and Modern Cultures*, 145.

9. Adams, "Ecofeminism and the Eating of Animals" (1996); Franklin, *Animals and Modern Cultures*; Rogers, "Beasts, Burgers, and Hummers."

10. Noske, *Humans and Other Animals*. Twine, "Revealing the 'Animal Industrial Complex.'" Fitzgerald and Taylor, "Cultural Hegemony of Meat and the Animal Industrial Complex."

11. U.S. Department of Agriculture and U.S. Department of Health and Human Services, "Dietary Guidelines for Americans 2010."

12. Franklin, *Animals and Modern Cultures*.

13. Horowitz, *Putting Meat on the American Table*; Franklin, *Animals and Modern Cultures*.

14. Dauvergne, *Shadows of Consumption*; Franklin, *Animals and Modern Cultures*.

15. Dauvergne, *Shadows of Consumption*.

16. Horowitz, *Putting Meat on the American Table*.

17. Horowitz, *Putting Meat on the American Table*.

18. Franklin, *Animals and Modern Cultures*; Horowitz, *Putting Meat on the American Table*.

19. Horowitz, *Putting Meat on the American Table*.

20. See Horowitz, *Putting Meat on the American Table* for a more detailed breakdown of types of meats consumed.

21. Skaggs, *Prime Cut*.

22. Horowitz, *Putting Meat on the American Table*; Skaggs, *Prime Cut*.

23. Skaggs, *Prime Cut*.

24. Horowitz, *Putting Meat on the American Table*.

25. Franklin, *Animals and Modern Cultures*; Horowitz, *Putting Meat on the American Table*.

26. Horowitz, *Putting Meat on the American Table*.

27. Franklin, *Animals and Modern Cultures*, 148.

28. Horowitz, *Putting Meat on the American Table*; Striffler, *Chicken*.

29. Earth Policy Institute, "Food and Agriculture"; USDA, Economic Research Service, "Livestock and Meat Domestic Data."

30. Wilkie, "Sentient Commodities and Productive Paradoxes."

31. Horowitz, *Putting Meat on the American Table*.

32. Earth Policy Institute, "Food and Agriculture"; USDA, Economic Research Service, "Livestock and Meat Domestic Data."

33. Earth Policy Institute, "Food and Agriculture"; USDA, Economic Research Service, "Livestock and Meat Domestic Data."

34. West and Fenstermaker, "Power, Inequality, and the Accomplishment of Gender."

35. Expenditures on fast food have continued to climb, increasing from approximately $6 billion in 1970 to $110 billion at the turn of the twenty-first century (Schlosser, *Fast Food Nation*).

36. Stull and Broadway, *Slaughterhouse Blues*.

37. Franklin, *Animals and Modern Cultures*, 154.

38. Adams, "Ecofeminism and the Eating of Animals" (1996); Rogers, "Beasts, Burgers, and Hummers; Rothgerber, "Real Men Don't Eat (Vegetable) Quiche"; Franklin, *Animals and Modern Cultures*.

39. Ruby, "Vegetarianism."

40. Rothgerber, "Real Men Don't Eat (Vegetable) Quiche."

41. Parry, "Gender and Slaughter in Popular Gastronomy."

42. Rogers, "Beasts, Burgers, and Hummers."

43. Gossard and York, "Social Structural Influences on Meat Consumption."

44. Kendall, Lobao, and Sharp, "Public Concern with Animal Well-Being," 423.

45. Rothgerber, "Real Men Don't Eat (Vegetable) Quiche," 263.

46. Ruby, "Vegetarianism."

47. Horowitz, *Putting Meat on the American Table*, 152.

48. Franklin, *Animals and Modern Cultures*; Gossard and York, "Social Structural Influences on Meat Consumption."

49. Franklin, *Animals and Modern Cultures*. It should be noted that beef consumption is an anomaly:

it increases with income, which is likely due to the higher cost (Gossard and York, "Social Structural Influences on Meat Consumption").

50. Gossard and York, "Social Structural Influences on Meat Consumption."
51. Gossard and York, "Social Structural Influences on Meat Consumption."
52. Gossard and York, "Social Structural Influences on Meat Consumption," 7.
53. Earth Policy Institute, "Food and Agriculture"; USDA, Economic Research Service, "Livestock and Meat Domestic Data."
54. Earth Policy Institute, "Food and Agriculture"; USDA, Economic Research Service, "Livestock and Meat Domestic Data."
55. Nierenberg, "Factory Farming in the Developing World."
56. Delgado, "Rising Consumption of Meat and Milk in Developing Countries."
57. FAO, "Livestock Industrialization, Trade and Social-Health-Environmental Impacts in Developing Countries"; Delgado, "Rising Consumption of Meat and Milk in Developing Countries."
58. FAO, "State of Food and Agriculture."
59. Marcus, *Meat Market*.
60. Delgado, "Rising Consumption of Meat and Milk in Developing Countries," 3910S.
61. Franklin, *Animals and Modern Cultures*, 146.
62. Franklin, *Animals and Modern Cultures*.
63. Elias, *The Civilizing Process*.
64. Rifkin, *Beyond Beef*.
65. Quoted in Singer and Mason, *The Way We Eat*, 11–12.
66. Gouveia and Juska, "Taming Nature, Taming Workers."
67. See Pachirat, *Every Twelve Seconds*.
68. Gouveia and Juska, "Taming Nature, Taming Workers," 372.
69. Franklin, *Animals and Modern Cultures*; Shukin, *Animal Capital*.
70. Franklin, *Animals and Modern Cultures*, 154.
71. Fitzgerald and Taylor, "Cultural Hegemony of Meat and the Animal Industrial Complex."
72. Glenn, "Constructing Consumables and Consent"; Fitzgerald and Taylor, "Cultural Hegemony of Meat and the Animal Industrial Complex."
73. Cole, "From 'Animal Machines' to 'Happy Meat'?"; Fitzgerald and Taylor, "Cultural Hegemony of Meat and the Animal Industrial Complex."
74. Croney, "Words Matter."
75. Heinz and Lee, "Getting Down to the Meat."
76. Dunayer, *Animal Equality*, 126.
77. Dunayer, *Animal Equality*.
78. Arluke and Sanders, *Regarding Animals*.
79. Adams, "Ecofeminism and the Eating of Animals" (1996).
80. Dunayer, *Animal Equality*.
81. Croney, "Words Matter," 102.
82. Franklin, *Animals and Modern Cultures*.
83. Williams, "Affected Ignorance and Animal Suffering."
84. Stuart and Worosz, "Risk, Anti-reflexivity, and Ethical Neutralization in Industrial Food Processing."
85. Dunayer, *Animal Equality*, 138.
86. Dunayer, *Animal Equality*.

87. Scully, *Dominion*; Kalechofsky, *Judaism and Animal Rights*; Foltz, *Animals in Islamic Tradition and Muslim Cultures*; Phelps, *The Great Compassion*; S. Rosen, *Holy Cow*.
88. Gouveia and Juska, "Taming Nature, Taming Workers."
89. Festinger, *A Theory of Cognitive Dissonance*.
90. Mayfield et al., "Consumption of Welfare-Friendly Food Products," 63.
91. Kendall, Lobao, and Sharp, "Public Concern with Animal Well-Being."
92. Kalof et al., "Social Psychological and Structural Influences on Vegetarian Beliefs."
93. Kendall, Lobao, and Sharp, "Public Concern with Animal Well-Being"; Kalof et al., "Social Psychological and Structural Influences on Vegetarian Beliefs"; Mertig and Dunlap, "Environmentalism, New Social Movements, and the New Class."
94. Bulliet, *Hunters, Herders, and Hamburgers*, 3.
95. Bulliet, *Hunters, Herders, and Hamburgers*, 182.
96. Herzog, *Some We Love, Some We Hate, Some We Eat*.
97. Serpell, "Having Our Dogs and Eating Them Too."
98. Meens, "Eating Animals in the Early Middle Ages"; Sorenson, *About Canada*.
99. Sorenson, *About Canada*.
100. Serpell, "Having Our Dogs and Eating Them Too"; Meens, "Eating Animals in the Early Middle Ages"; Sorenson, *About Canada*.
101. Serpell, *In the Company of Animals*.
102. Serpell, *In the Company of Animals*.
103. Dunayer, *Animal Equality*.
104. Meens, "Eating Animals in the Early Middle Ages."
105. Dunayer, *Animal Equality*.
106. Francione, *Introduction to Animal Rights*; Serpell, "Having Our Dogs and Eating Them Too."
107. Serpell, *In the Company of Animals*, 18.
108. Croney, "Words Matter."
109. Serpell, "Having Our Dogs and Eating Them Too," 642.
110. Serpell, "Having Our Dogs and Eating Them Too," 642.
111. Serpell, *In the Company of Animals*.
112. Serpell, *In the Company of Animals*, 235–36.

CHAPTER 5. INDUSTRIALIZATION FALLOUT

1. D. Fitzgerald, *Every Farm a Factory*.
2. Weis, "Accelerating Biophysical Contradictions of Industrial Capitalist Agriculture"; Imhoff, "Introduction."
3. C. Adams, "Ecofeminism and the Eating of Animals" (1991), 129.
4. Rollin, "Farm Factories."
5. Matheny and Leahy, "Farm-Animal Welfare, Legislation, and Trade," 328.
6. Stathopoulos, "You Are What Your Food Eats."
7. Imhoff, "Introduction," xii.
8. Rollin, "Farm Factories."
9. D'Silva, "Adverse Impact of Industrial Animal Agriculture on the Health and Welfare of Farmed Animals"; Mench et al., "Welfare of Animals in Concentrated Animal Feeding Operations."

10. D'Silva, "Adverse Impact of Industrial Animal Agriculture on the Health and Welfare of Farmed Animals."

11. Mench et al., "Welfare of Animals in Concentrated Animal Feeding Operations"; Mason and Finelli, "Brave New Farm?"

12. Mench et al., "Welfare of Animals in Concentrated Animal Feeding Operations."

13. Stathopoulos, "You Are What Your Food Eats."

14. Mench et al., "Welfare of Animals in Concentrated Animal Feeding Operations."

15. Stathopoulos, "You Are What Your Food Eats," 416.

16. Mench et al., "Welfare of Animals in Concentrated Animal Feeding Operations."

17. Mench et al., "Welfare of Animals in Concentrated Animal Feeding Operations."

18. Eisnitz, *Slaughterhouse*; Warrick, "'They Die Piece by Piece.'"

19. Ransom, "Rise of Agricultural Animal Welfare Standards."

20. Isaacs-Blundin, "Why Manure May Be the Farm Animal Advocate's Best Friend"; Vining, "Animal Cruelty Laws and Factory Farming"; Cassuto, "Bred Meat"; Mason and Finelli, "Brave New Farm?"; Matheny and Leahy, "Farm-Animal Welfare, Legislation, and Trade"; Mallon, "Deplorable Standard of Living Faced by Farmed Animals in America's Meat Industry."

21. Matheny and Leahy, "Farm-Animal Welfare, Legislation, and Trade," 337.

22. U.S. Department of Labor, Bureau of Labor Statistics, "Occupational Employment and Wages."

23. Striffler, *Chicken*.

24. U.S. Department of Labor, Bureau of Labor Statistics, "Occupational Employment and Wages."

25. HHS (U.S. Department of Health and Human Services), "2012 HHS Poverty Guidelines."

26. U.S. Department of Labor, Bureau of Labor Statistics, "Table SNR05."

27. Striffler, *Chicken*.

28. For an engaging account of the plight of (primarily immigrant) workers in the industry, see Schlosser, *Fast Food Nation*.

29. Schlosser, *Fast Food Nation*; Tietz, "Boss Hog."

30. Donham et al., "Community Health and Socioeconomic Issues Surrounding Concentrated Animal Feeding Operations."

31. Benson, "Effects of Packinghouse Work on Southeast Asian Refugee Families"; Broadway and Stull, "'I'll Do Whatever You Want, but It Hurts'"; Olsson, "The Shame of Meatpacking"; Stull, "Knock 'Em Dead"; Eisnitz, *Slaughterhouse*; Striffler, "Watching the Chickens Pass By."

32. Cook, "Sliced and Diced."

33. Broadway and Stull, "Meat Processing and Garden City, KS."

34. Joy, *Why We Love Dogs, Eat Pigs and Wear Cows*, 80.

35. Cook, "Sliced and Diced."

36. Broadway, "Planning for Change in Small Towns or Trying to Avoid the Slaughterhouse Blues."

37. Cook, "Sliced and Diced," 233.

38. Hemsworth and Coleman, *Human-Livestock Interactions*.

39. Noske, *Humans and Other Animals*.

40. *New York Times*, "What Meat Means."

41. Eisnitz, *Slaughterhouse*; Fink, *Cutting into the Meatpacking Line*; Rémy, "Une mise à mort industrielle 'humaine'?"

42. Fink, *Cutting into the Meatpacking Line*, 37.

43. Joy, *Why We Love Dogs, Eat Pigs and Wear Cows*.

44. Associated Press, "Two Slaughterhouse Workers Charged with Abuse."

45. Herzog and McGee, "Psychological Aspects of Slaughter."

46. Emhan et al., "Psychological Symptom Profile of Butchers Working in Slaughterhouse and Retail Meat Packing Business."

47. MacNair, *Perpetration-Induced Traumatic Stress*, 88.

48. Mallon, "Deplorable Standard of Living Faced by Farmed Animals in America's Meat Industry."

49. See, for instance, the comprehensive study by Campbell and Campbell, *The China Study*.

50. Stathopoulos, "You Are What Your Food Eats."

51. Horrigan, Lawrence, and Walker, "How Sustainable Agriculture Can Address the Environmental and Health Harms of Industrial Agriculture."

52. Mallon, "Deplorable Standard of Living Faced by Farmed Animals in America's Meat Industry," 389–90.

53. CDC, "Estimates of Foodborne Illness in the United States."

54. CDC, "CDC 2011 Estimates: Findings."

55. Mason and Finelli, "Brave New Farm?"

56. CDC, "Vital Signs."

57. Juska et al., "Manufacturing Bacterological Contamination Outbreaks in Industrialized Meat Production Systems"; Stathopoulos, "You Are What Your Food Eats."

58. Juska et al., "Manufacturing Bacteriological Contamination Outbreaks in Industrialized Meat Production Systems."

59. Mallon, "Deplorable Standard of Living Faced by Farmed Animals in America's Meat Industry."

60. Stathopoulos, "You Are What Your Food Eats."

61. Gouveia and Juska, "Taming Nature, Taming Workers."

62. Juska et al., "Manufacturing Bacteriological Contamination Outbreaks in Industrialized Meat Production Systems."

63. Juska et al., "Manufacturing Bacteriological Contamination Outbreaks in Industrialized Meat Production Systems."

64. Joy, *Why We Love Dogs, Eat Pigs and Wear Cows*, 79.

65. CDC, "Prion Diseases."

66. Mason and Finelli, "Brave New Farm?"

67. Shukin, *Animal Capital*.

68. Stathopoulos, "You Are What Your Food Eats."

69. Pollan, "Power Steer," 108.

70. Pollan, "Power Steer"; Stathopoulos, "You Are What Your Food Eats."

71. Stathopoulos, "You Are What Your Food Eats."

72. A. Fitzgerald, "The 'Underdog' as 'Ideal Victim'?"

73. Mallon, "Deplorable Standard of Living Faced by Farmed Animals in America's Meat Industry."

74. Stathopoulos, "You Are What Your Food Eats."

75. Shukin, *Animal Capital*.

76. Schmidt, "Swine CAFOs and Novel H1N1 Flu"; Gilchrist et al., "Potential Role of Concentrated Animal Feeding Operations in Infectious Disease Epidemics and Antibiotic Resistance"; Saenz, Hethcote, and Gray, "Confined Animal Feeding Operations as Amplifiers of Influenza"; Stathopoulos, "You Are What Your Food Eats."

77. Shukin, *Animal Capital*; Saenz, Hethcote, and Gray, "Confined Animal Feeding Operations as Amplifiers of Influenza."

78. Stathopoulos, "You Are What Your Food Eats."

79. Mason and Finelli, "Brave New Farm?"

80. Schmidt, "Swine CAFOs and Novel H1N1 Flu"; Stathopoulos, "You Are What Your Food Eats."

81. CDC, "Updated CDC Estimates of 2009 H1N1 Influenza Cases, Hospitalizations and Deaths in the United States."

82. Schmidt, "Swine CAFOs and Novel H1N1 Flu."

83. Saenz, Hethcote, and Gray, "Confined Animal Feeding Operations as Amplifiers of Influenza," 341.

84. Saenz, Hethcote, and Gray. "Confined Animal Feeding Operations as Amplifiers of Influenza."

85. Schmidt, "Swine CAFOs and Novel H1N1 Flu."

86. Schmidt, "Swine CAFOs and Novel H1N1 Flu," A400–A401.

87. Mason and Finelli, "Brave New Farm?"; Stathopoulos, "You Are What Your Food Eats."

88. Gilchrist et al., "Potential Role of Concentrated Animal Feeding Operations in Infectious Disease Epidemics and Antibiotic Resistance."

89. Dauvergne, *Shadows of Consumption*.

90. Gilchrist et al., "Potential Role of Concentrated Animal Feeding Operations in Infectious Disease Epidemics and Antibiotic Resistance."

91. Mason and Finelli, "Brave New Farm?"

92. Green et al., "Bacterial Plume Emanating from the Air Surrounding Swine Confinement Operations."

93. Gilchrist et al., "Potential Role of Concentrated Animal Feeding Operations in Infectious Disease Epidemics and Antibiotic Resistance."

94. Stathopoulos, "You Are What Your Food Eats."

95. Green et al., "Bacterial Plume Emanating from the Air Surrounding Swine Confinement Operations."

96. Imhoff, "Introduction."

97. Gilchrist et al., "Potential Role of Concentrated Animal Feeding Operations in Infectious Disease Epidemics and Antibiotic Resistance," 313.

98. Gilchrist et al., "Potential Role of Concentrated Animal Feeding Operations in Infectious Disease Epidemics and Antibiotic Resistance."

99. Mallon, "Deplorable Standard of Living Faced by Farmed Animals in America's Meat Industry."

100. Hahn Niman, *Righteous Porkchop*.

101. Packwood Freeman, "Meat's Place on the Campaign Menu," 256.

102. Weis, "Accelerating Biophysical Contradictions of Industrial Capitalist Agriculture."

103. Mirabelli et al., "Asthma Symptoms among Adolescents Who Attend Public Schools That Are Located Near Confined Swine Feeding Operations"; Donham et al., "Community Health and Socioeconomic Issues Surrounding Concentrated Animal Feeding Operations."

104. Mirabelli et al., "Asthma Symptoms among Adolescents Who Attend Public Schools That Are Located Near Confined Swine Feeding Operations."

105. Tietz, "Boss Hog."

106. Donham et al., "Community Health and Socioeconomic Issues Surrounding Concentrated Animal Feeding Operations."

107. Green et al., "Bacterial Plume Emanating from the Air Surrounding Swine Confinement Operations."

108. Zande, "Raising a Stink."

109. Centner, "Governmental Oversight of Discharges from Concentrated Animal Feeding Operations."

110. Hahn Niman, *Righteous Porkchop*; Tietz, "Boss Hog"; Stathopoulos, "You Are What Your Food Eats."

111. Stathopoulos, "You Are What Your Food Eats"; Hahn Niman, *Righteous Porkchop*.
112. U.S. General Accounting Office, "Animal Agriculture."
113. Centner, "Governmental Oversight of Discharges from Concentrated Animal Feeding Operations."
114. Tietz, "Boss Hog," 111.
115. Isaacs-Blundin, "Why Manure May Be the Farm Animal Advocate's Best Friend."
116. Hahn Niman, *Righteous Porkchop*; Zande, "Raising a Stink."
117. Hahn Niman, *Righteous Porkchop*.
118. Tietz, "Boss Hog."
119. West et al., "Antibiotic Resistance, Gene Transfer, and Water Quality Patterns"; Centner, "Governmental Oversight of Discharges from Concentrated Animal Feeding Operations."
120. Zande, "Raising a Stink."
121. West et al., "Antibiotic Resistance, Gene Transfer, and Water Quality Patterns."
122. Hahn Niman, *Righteous Porkchop*.
123. Tietz, "Boss Hog."
124. Steinfeld et al., "Livestock's Long Shadow."
125. Packwood Freeman, "Meat's Place on the Campaign Menu."
126. Lappé, "Diet for a Hot Planet."
127. Steinfeld et al., "Livestock's Long Shadow."
128. Gunderson, "Metabolic Rifts of Livestock Agribusiness."
129. Steinfeld et al., "Livestock's Long Shadow."
130. Bristow and Fitzgerald, "Global Climate Change and the Industrial Animal Agriculture Link."
131. Mallon, "Deplorable Standard of Living Faced by Farmed Animals in America's Meat Industry."
132. Ponette-González and Fry, "Pig Pandemic"; Nibert, *Animal Rights/Human Rights*.
133. Winders and Nibert, "Consuming the Surplus," 90.
134. Winders and Nibert, "Consuming the Surplus," 90.
135. Dauvergne, *Shadows of Consumption*; Imhoff, "Introduction."
136. Winders and Nibert, "Consuming the Surplus."
137. Dauvergne, *Shadows of Consumption*.
138. Nierenberg, "Factory Farming in the Developing World."
139. Rothgerber, "Real Men Don't Eat (Vegetable) Quiche."
140. Marcus, *Meat Market*, 226.
141. Goldschmidt, *As You Sow*.
142. Artz, Orazem, and Otto, "Measuring the Impact of Meat Packing and Processing Facilities in Nonmetropolitan Counties."
143. *National Farmers Union News*, "Farmers Union Members Testify on Negative Impacts of Livestock Concentration."
144. Noble, "Paying the Polluters."
145. Kilpatrick, "Concentrated Animal Feeding Operations and Proximate Property Values."
146. Imhoff, "Introduction."
147. Noble, "Paying the Polluters," 231.
148. Schmidt, "Low on the Hog," A134.
149. Donham et al., "Community Health and Socioeconomic Issues Surrounding Concentrated Animal Feeding Operations."
150. Donham et al., "Community Health and Socioeconomic Issues Surrounding Concentrated Animal Feeding Operations"; Schmidt, "Low on the Hog."

151. Broadway, "Planning for Change in Small Towns or Trying to Avoid the Slaughterhouse Blues"; Stull and Broadway, *Slaughterhouse Blues*.

152. Broadway, "Planning for Change in Small Towns or Trying to Avoid the Slaughterhouse Blues."

153. Broadway, "Planning for Change in Small Towns or Trying to Avoid the Slaughterhouse Blues"; Grey, "Meatpacking in Storm Lake, Iowa"; Stull and Broadway, *Slaughterhouse Blues*; Grey, "Pork, Poultry and Newcomers in Storm Lake, Iowa"; Horowitz and Miller, *Immigrants in the Delmarva Poultry Processing Industry*.

154. A. Fitzgerald, "A Social History of the Slaughterhouse."

155. Gouveia and Juska, "Taming Nature, Taming Workers"; Bacon, "INS Declares War on Labor"; Dalla, Ellis, and Cramer, "Immigration and Rural America"; Grey, "Meatpacking in Storm Lake, Iowa."

156. Broadway, "Bad to the Bone"; Broadway, "Meatpacking and Its Social and Economic Consequences for Garden City, Kansas in the 1980s"; Broadway, "Planning for Change in Small Towns or Trying to Avoid the Slaughterhouse Blues"; Broadway, "What Happens When the Meatpackers Come to Town?"; Stull and Broadway, *Slaughterhouse Blues*.

157. Broadway, "Meatpacking and the Transformation of Rural Communities"; Broadway, "Planning for Change in Small Towns or Trying to Avoid the Slaughterhouse Blues"; Marcus, *Meat Market*; Stull and Broadway, *Slaughterhouse Blues*.

158. Eisnitz, *Slaughterhouse*; Schlosser, *Fast Food Nation*.

159. Fitzgerald, Kalof, and Dietz, "Slaughterhouses and Increased Crime Rates."

160. Sinclair, *The Jungle*, 18–19 (emphasis mine).

161. Fitzgerald, Kalof, and Dietz, "Slaughterhouses and Increased Crime Rates."

162. Artz, Orazem, and Otto, "Measuring the Impact of Meat Packing and Processing Facilities in Nonmetropolitan Counties," 568.

163. Allen, "Influence of Military Training and Combat Experience on Domestic Violence"; Marshall, Panuzio, and Taft, "Intimate Partner Violence among Military Veterans and Active Duty Servicemen"; Marshall and McShane, "First to Fight"; Mercier, "Violence in the Military Family"; Rosen et al., "Effects of Peer Group Climate on Intimate Partner Violence among Married Male U.S. Army Soldiers"; Black, "Stress and the Correctional Officer"; Kauffman, *Prison Officers and Their World*; Stack and Tsoudis, "Suicide Risk among Correctional Officers."

164. Cochran and Chamlin, "Deterrence and Brutalization"; Cochran, Chamlin, and Seth, "Deterence or Brutalization?"; King, "The Brutalization Effect"; Yang and Lester, "Deterrent Effect of Executions."

165. Hardeman and Jochemsen, "Are There Ideological Aspects to the Modernization of Agriculture?"

166. Kimbrell, "Cold Evil."

167. Thu, "CAFOs Are in Everyone's Backyard"; Joy, *Why We Love Dogs, Eat Pigs and Wear Cows*, 91.

168. Thu, "CAFOs Are in Everyone's Backyard," 220.

169. Ibrahim, "A Return to Descartes," 97.

170. Fitzgerald and Taylor, "The Cultural Hegemony of Meat and the Animal Industrial Complex."

171. Bulliet, *Hunters, Herders, and Hamburgers*.

172. Bulliet, *Hunters, Herders, and Hamburgers*, 3.

173. Otter, "Civilizing Slaughter," 105.

174. Joy, *Why We Love Dogs, Eat Pigs and Wear Cows*, 71; Williams, "Affected Ignorance and Animal Suffering."

175. Bulliet, *Hunters, Herders, and Hamburgers*.

CHAPTER 6. BRIDGING THE DIVIDE BETWEEN PRODUCTION, PROCESSING, CONSUMPTION, AND IMPACTS

1. Horrigan, Lawrence, and Walker, "How Sustainable Agriculture Can Address the Environmental and Health Harms of Industrial Agriculture."
2. Cassuto, "Bred Meat," 87.
3. Scully, "Fear Factories," 22.
4. R. Adams, "Fast Food and Animal Rights"; Jasper and Nelkin, *Animal Rights Crusade*.
5. Francione, *Rain without Thunder*.
6. Francione, *Rain without Thunder*.
7. Matheny and Leahy, "Farm-Animal Welfare, Legislation, and Trade."
8. Vanhonacker et al., "Segmentation Based on Consumers' Perceived Importance and Attitude toward Farm Animal Welfare."
9. Bulliet, *Hunters, Herders, and Hamburgers*.
10. Ransom, "Rise of Agricultural Animal Welfare Standards."
11. Matheny and Leahy, "Farm-Animal Welfare, Legislation, and Trade."
12. Matheny and Leahy, "Farm-Animal Welfare, Legislation, and Trade."
13. Marcus, *Meat Market*.
14. Pollan, *Omnivore's Dilemma*.
15. Twine, "Searching for the 'Win-Win'?"
16. Mench et al., "Welfare of Animals in Concentrated Animal Feeding Operations."
17. Twine, "Searching for the 'Win-Win'?"
18. Ibrahim, "A Return to Descartes"; R. Adams, "Fast Food and Animal Rights."
19. R. Adams, "Fast Food and Animal Rights," 323.
20. Zogby, Bruce, and Wittman, "Nationwide Views on the Treatment of Farm Animals."
21. Matheny and Leahy, "Farm-Animal Welfare, Legislation, and Trade."
22. Weis, "Accelerating Biophysical Contradictions of Industrial Capitalist Agriculture"; Matheny and Leahy, "Farm-Animal Welfare, Legislation, and Trade."
23. Bristow and Fitzgerald, "Global Climate Change and the Industrial Animal Agriculture Link."
24. Gossard and York, "Social Structural Influences on Meat Consumption."
25. Mayfield et al., "Consumption of Welfare-Friendly Food Products"; Barnard, "World Cancer Day."
26. Mench et al., "Welfare of Animals in Concentrated Animal Feeding Operations."
27. Mench et al., "Welfare of Animals in Concentrated Animal Feeding Operations"; Ibrahim, "A Return to Descartes."
28. Bjerklie, "Size Matters."
29. Mench et al., "Welfare of Animals in Concentrated Animal Feeding Operations"; Matheny and Leahy, "Farm-Animal Welfare, Legislation, and Trade."
30. Mench et al., "Welfare of Animals in Concentrated Animal Feeding Operations."
31. Matheny and Leahy, "Farm-Animal Welfare, Legislation, and Trade," 355.
32. Matheny and Leahy, "Farm-Animal Welfare, Legislation, and Trade"; Mench et al., "Welfare of Animals in Concentrated Animal Feeding Operations."
33. R. Adams, "Fast Food and Animal Rights."
34. Mench et al., "Welfare of Animals in Concentrated Animal Feeding Operations"; Ransom, "Rise of Agricultural Animal Welfare Standards."
35. Bjerklie, "Size Matters."

36. R. Adams, "Fast Food and Animal Rights," 317.

37. R. Adams, "Fast Food and Animal Rights"; Ransom, "Rise of Agricultural Animal Welfare Standards."

38. Matheny and Leahy, "Farm-Animal Welfare, Legislation, and Trade," 356.

39. Ibrahim, "A Return to Descartes."

40. Ibrahim, "A Return to Descartes"; Ransom, "Rise of Agricultural Animal Welfare Standards."

41. Horrigan, Lawrence, and Walker, "How Sustainable Agriculture Can Address the Environmental and Health Harms of Industrial Agriculture," 453.

42. Horrigan, Lawrence, and Walker, "How Sustainable Agriculture Can Address the Environmental and Health Harms of Industrial Agriculture." See also Cotton, *Overshoot*.

43. Weis, "Accelerating Biophysical Contradictions of Industrial Capitalist Agriculture"; Horrigan, Lawrence, and Walker, "How Sustainable Agriculture Can Address the Environmental and Health Harms of Industrial Agriculture."

44. Horrigan, Lawrence, and Walker, "How Sustainable Agriculture Can Address the Environmental and Health Harms of Industrial Agriculture."

45. Greene et al., "Emerging Issues in the US Organic Industry."

46. Horrigan, Lawrence, and Walker, "How Sustainable Agriculture Can Address the Environmental and Health Harms of Industrial Agriculture."

47. Stuart and Worosz, "Risk, Anti-reflexivity, and Ethical Neutralization in Industrial Food Processing."

48. Horrigan, Lawrence, and Walker, "How Sustainable Agriculture Can Address the Environmental and Health Harms of Industrial Agriculture."

49. Mench et al., "Welfare of Animals in Concentrated Animal Feeding Operations"; Dauvergne, *Shadows of Consumption*.

50. Gunderson, "Metabolic Rifts of Livestock Agribusiness"; Dauvergne, *Shadows of Consumption*; Ibrahim, "A Return to Descartes."

51. Weis, "Accelerating Biophysical Contradictions of Industrial Capitalist Agriculture," 334 (emphasis mine).

52. Mench et al., "Welfare of Animals in Concentrated Animal Feeding Operations."

53. Zogby, Bruce, and Wittman, "Nationwide Views on the Treatment of Farm Animals."

54. Mench et al., "Welfare of Animals in Concentrated Animal Feeding Operations."

55. Mallon, "Deplorable Standard of Living Faced by Farmed Animals in America's Meat Industry"; Cassuto, "Bred Meat."

56. Mench et al., "Welfare of Animals in Concentrated Animal Feeding Operations."

57. Mallon, "Deplorable Standard of Living Faced by Farmed Animals in America's Meat Industry."

58. Vining, "Animal Cruelty Laws and Factory Farming," 123.

59. Mench et al., "Welfare of Animals in Concentrated Animal Feeding Operations."

60. Cassuto, "Bred Meat."

61. Snider, "Cooperative Models and Corporate Crime."

62. Matheny and Leahy, "Farm-Animal Welfare, Legislation, and Trade."

63. Cassuto, "Bred Meat"; Francione, "Animals, Property, and Personhood"; Mallon, "Deplorable Standard of Living Faced by Farmed Animals in America's Meat Industry."

64. Cassuto, "Bred Meat."

65. Horrigan, Lawrence, and Walker, "How Sustainable Agriculture Can Address the Environmental and Health Harms of Industrial Agriculture," 454.

66. Gilchrist et al., "Potential Role of Concentrated Animal Feeding Operations in Infectious Disease

Epidemics and Antibiotic Resistance"; Zande, "Raising a Stink"; Greene et al., "Emerging Issues in the US Organic Industry."

67. Gilchrist et al., "Potential Role of Concentrated Animal Feeding Operations in Infectious Disease Epidemics and Antibiotic Resistance"; Zande, "Raising a Stink."

68. Donham et al., "Community Health and Socioeconomic Issues Surrounding Concentrated Animal Feeding Operations."

69. Donham et al., "Community Health and Socioeconomic Issues Surrounding Concentrated Animal Feeding Operations."

70. Zande, "Raising a Stink"; Tietz, "Boss Hog."

71. Stuart and Worosz. "Myth of Efficiency."

72. Hahn Niman, *Righteous Porkchop*.

73. Packwood Freeman, "Meat's Place on the Campaign Menu"; Matheny and Leahy, "Farm-Animal Welfare, Legislation, and Trade."

74. See Matheny and Leahy, "Farm-Animal Welfare, Legislation, and Trade" for a discussion of the possible articles of the General Agreement on Tariffs and Trade (GATT) that could apply to restrictions on trade in animal products.

75. Mench et al., "Welfare of Animals in Concentrated Animal Feeding Operations."

76. Mench et al., "Welfare of Animals in Concentrated Animal Feeding Operations"; Matheny and Leahy, "Farm-Animal Welfare, Legislation, and Trade."

77. Stuart and Worosz, "Myth of Efficiency"; Snider, "Cooperative Models and Corporate Crime"; Kirkhorne and Schenker, "Human Health Effects of Agriculture"; Centner, "Governmental Oversight of Discharges from Concentrated Animal Feeding Operations."

78. Mallon, "Deplorable Standard of Living Faced by Farmed Animals in America's Meat Industry," 410.

79. Zande, "Raising a Stink"; Donham et al., "Community Health and Socioeconomic Issues Surrounding Concentrated Animal Feeding Operations."

80. Stathopoulos, "You Are What Your Food Eats."

81. Schlosser, "Bad Meat"; Mallon, "Deplorable Standard of Living Faced by Farmed Animals in America's Meat Industry."

82. Mallon, "Deplorable Standard of Living Faced by Farmed Animals in America's Meat Industry"; Matheny and Leahy, "Farm-Animal Welfare, Legislation, and Trade."

83. Cassuto, "Bred Meat."

84. Horrigan, Lawrence, and Walker, "How Sustainable Agriculture Can Address the Environmental and Health Harms of Industrial Agriculture."

85. Starmer, Witteman, and Wise, "Feeding the Factory Farm"; Horrigan, Lawrence, and Walker, "How Sustainable Agriculture Can Address the Environmental and Health Harms of Industrial Agriculture"; Weis, "Accelerating Biophysical Contradictions of Industrial Capitalist Agriculture."

86. Horrigan, Lawrence, and Walker, "How Sustainable Agriculture Can Address the Environmental and Health Harms of Industrial Agriculture," 453.

87. Starmer, Witteman, and Wise, "Feeding the Factory Farm"; Mench et al., "Welfare of Animals in Concentrated Animal Feeding Operations."

88. Cited by Schmidt, "Swine CAFOs and Novel H1N1 Flu," A401.

89. Horrigan, Lawrence, and Walker, "How Sustainable Agriculture Can Address the Environmental and Health Harms of Industrial Agriculture."

90. Horrigan, Lawrence, and Walker, "How Sustainable Agriculture Can Address the Environmental and Health Harms of Industrial Agriculture"; Donham et al., "Community Health and Socioeconomic Issues Surrounding Concentrated Animal Feeding Operations."

91. Francione, "Animals, Property, and Personhood."

92. Ransom, "Rise of Agricultural Animal Welfare Standards"; World Organisation for Animal Health, "OIE's Achievements in Animal Welfare."

93. Hardeman and Jochemsen, "Are There Ideological Aspects to the Modernization of Agriculture?"; Matheny and Leahy, "Farm-Animal Welfare, Legislation, and Trade"; Twine, *Animals as Biotechnology*.

94. Matheny and Leahy, "Farm-Animal Welfare, Legislation, and Trade."

95. Cook, "Sliced and Diced."

96. Striffler, "Watching the Chickens Pass By."

97. Mench et al., "Welfare of Animals in Concentrated Animal Feeding Operations"; Hemsworth and Coleman, *Human-Livestock Interactions*; Wilkie, "Sentient Commodities and Productive Paradoxes"; Saenz, Hethcote, and Gray, "Confined Animal Feeding Operations as Amplifiers of Influenza"; Green et al., "Bacterial Plume Emanating from the Air Surrounding Swine Confinement Operations."

98. Fitzgerald, Kalof, and Dietz, "Slaughterhouses and Increased Crime Rates"; MacNair, *Perpetration-Induced Traumatic Stress*.

99. Dillard, "A Slaughterhouse Nightmare."

100. Gilchrist et al., "Potential Role of Concentrated Animal Feeding Operations in Infectious Disease Epidemics and Antibiotic Resistance"; Stathopoulos, "You Are What Your Food Eats."

101. Stathopoulos, "You Are What Your Food Eats."

102. Eisnitz, *Slaughterhouse*; Schlosser, *Fast Food Nation*.

103. Juska et al., "Manufacturing Bacteriological Contamination Outbreaks in Industrialized Meat Production Systems."

104. Stathopoulos, "You Are What Your Food Eats"; Gilchrist et al., "Potential Role of Concentrated Animal Feeding Operations in Infectious Disease Epidemics and Antibiotic Resistance."

105. Weis, "Accelerating Biophysical Contradictions of Industrial Capitalist Agriculture"

106. Horrigan, Lawrence, and Walker, "How Sustainable Agriculture Can Address the Environmental and Health Harms of Industrial Agriculture."

107. Singer and Mason, *The Way We Eat*.

108. Francione, "Animals, Property and Legal Welfarism," 721; Gunderson, "From Cattle to Capital"; Twine, "Searching for the 'Win-Win'?"

109. Ibrahim, "A Return to Descartes," 113.

110. Regan, "Preface," 2–3.

111. Stănescu, "'Green' Eggs and Ham?"

112. Ruby, "Vegetarianism."

113. Franklin, *Animals and Modern Cultures*; Maurer, *Vegetarianism*.

114. Franklin, *Animals and Modern Cultures*.

115. Ruby, "Vegetarianism."

116. Maurer, *Vegetarianism*; Franklin, *Animals and Modern Cultures*.

117. Ruby, "Vegetarianism"; C. Adams, *Sexual Politics of Meat*.

118. Rothgerber, "Real Men Don't Eat (Vegetable) Quiche."

119. Maurer, *Vegetarianism*.

120. Packwood Freeman, "Meat's Place on the Campaign Menu"; Ruby, "Vegetarianism"; Rothgerber, "Real Men Don't Eat (Vegetable) Quiche."

121. Maurer, *Vegetarianism*.

122. Ruby, "Vegetarianism."

123. See for instance, C. Adams, "Ecofeminism and the Eating of Animals" (1991); Gaard, "Vegetarian Ecofeminism"; C. Adams, "Ecofeminism and the Eating of Animals" (1996).

124. Birke, "Intimate Familiarities?," 429.

125. Ruby, "Vegetarianism." See Maurer, *Vegetarianism* for a discussion of other reasons for abstaining from consuming meat.

126. Rothgerber, "Real Men Don't Eat (Vegetable) Quiche."

127. Ruby, "Vegetarianism."

128. Other demographic variables are also related to adopting a meat-free diet. In the United States, women, those with higher incomes and education, people with a White racial/ethnic background, residents of the Northeast and Western coast states, those in service and professional occupations, unreligious, liberal, and health conscious are more likely to abstain from eating meat (Ruby, "Vegetarianism"; Maurer, *Vegetarianism*).

129. Kalof et al., "Social Psychological and Structural Influences on Vegetarian Beliefs."

130. Ruby, "Vegetarianism."

131. Gouveia and Juska, "Taming Nature, Taming Workers"; Gunderson, "From Cattle to Capital"; and Gunderson, "Metabolic Rifts of Livestock Agribusiness."

132. Szasz, *Shopping Our Way to Safety* writes about this problem in the environmental movement. A parallel is clearly evident in the animal rights movement.

133. Gouveia and Juska, "Taming Nature, Taming Workers."

134. Pollan, *Omnivore's Dilemma*, 330.

135. Maurer, *Vegetarianism*.

136. Maurer, *Vegetarianism*.

137. Rothgerber, "Real Men Don't Eat (Vegetable) Quiche."

138. Gouveia and Juska, "Taming Nature, Taming Workers"; Maurer, *Vegetarianism*.

139. Marcus, *Meat Market*.

140. Hardeman and Jochemsen, "Are There Ideological Aspects to the Modernization of Agriculture?"

141. Weis, "Accelerating Biophysical Contradictions of Industrial Capitalist Agriculture," 334.

142. Gouveia and Juska, "Taming Nature, Taming Workers," 372.

143. Johnsen, *Raising a Stink*, xi.

144. Packwood Freeman, "Meat's Place on the Campaign Menu"; Bristow and Fitzgerald, "Global Climate Change and the Industrial Animal Agriculture Link."

145. Thu, "CAFOs Are in Everyone's Backyard."

146. Joy, *Why We Love Dogs, Eat Pigs and Wear Cows*, 142–43.

147. P. Adams, "First Lab-Grown Hamburger Is Served"; Feldstein, "Meet George Jetson—and His Test Tube Turkey?"

148. Gunderson, "From Cattle to Capital."

149. Weis, "Accelerating Biophysical Contradictions of Industrial Capitalist Agriculture," 318.

150. Shukin, *Animal Capital*, 232.

151. Edwards and Ladd, "Environmental Justice, Swine Production and Farm Loss in North Carolina"; Ladd and Edward, "Corporate Swine and Capitalist Pigs"; Tacquino, Parisi, and Gill, "Units of Analysis and the Environmental Justice Hypothesis"; Wing et al., "Integrating Epidemiology, Education, and Organizing for Environmental Justice."

152. A. Fitzgerald, "A Social History of the Slaughterhouse."

153. Weis, "Accelerating Biophysical Contradictions of Industrial Capitalist Agriculture," 337.

154. Bulliet, *Hunters, Herders, and Hamburgers*, 3.

155. Horowitz, *Putting Meat on the American Table*, 153.

156. A. Fitzgerald, "A Social History of the Slaughterhouse."
157. Gregory, *Animal Welfare and Meat Production*, 6–7.
158. *CBC News*, "Seal Product Ban Upheld on 'Ethical' Grounds."
159. Bulliet, *Hunters, Herders, and Hamburgers*.
160. Franklin, *Animals and Modern Cultures*, 174.
161. Pachirat, *Every Twelve Seconds*.
162. Thomas, *Man and the Natural World*, 303.

Bibliography

Abbott, Alison. "Pig Geneticists Go the Whole Hog: Genome Will Benefit Farmers and Medical Researchers." *Nature*, November 14, 2012. http://www.nature.com/news/pig-geneticists-go-the -whole-hog-1.11801.

Adams, Carol J. "Ecofeminism and the Eating of Animals." *Hypatia* 6, no. 1 (1991): 125–45.

———. "Ecofeminism and the Eating of Animals." In *Ecological Feminist Philosophies*, ed. by Karen J. Warren. Bloomington: Indiana University Press, 1996.

———. *The Sexual Politics of Meat: A Feminist-Vegetarian Critical Theory*. New York: Continuum, 1990.

Adams, Paul. "The First Lab-Grown Hamburger Is Served." *Popular Science*, August 5, 2013. http:// www.popsci.com/technology/article/2013-08/first-lab-grown-hamburger-served.

Adams, Ronald J. "Fast Food and Animal Rights: An Examination and Assessment of the Industry's Response to Social Pressure." *Business and Society Review* 113 (2008): 301–28.

Allen, Leana C. "The Influence of Military Training and Combat Experience on Domestic Violence." In *Battle Cries on the Home Front: Violence in the Military Family*, ed. by Peter J. Mercier and Judith D. Mercier. Springfield, IL: Charles C. Thomas, 2000.

Arluke, Arnold, and Clinton R. Sanders. *Regarding Animals*. Philadelphia: Temple University Press, 1996.

Artz, Georgeanne M., Peter F. Orazem, and Daniel M. Otto. "Measuring the Impact of Meat Packing and Processing Facilities in Nonmetropolitan Counties: A Difference-in-Differences Approach." *American Journal of Agricultural Economics* 89 (2007): 557–70.

Associated Press. "Two Slaughterhouse Workers Charged with Abuse." *USA Today*, February 17, 2008.

Atwood, Margaret. *The Edible Woman*. Toronto: McClelland and Stewart, 1969.

Azzam, Azzeddine M. "Competition in the US Meatpacking Industry: Is It History?" *Agricultural Economics* 18 (1998): 107–26.

Babad, Michael. "Apple's Shares Tumble, with Bulls 'Licking Their Wounds.'" *Globe and Mail*, January 24, 2013.

Bacon, David. "INS Declares War on Labor: Ethnic Cleansing Hits Immigrant Workers, Organizers, in Midwest Meatpacking." *The Nation* 268, no. 13 (1999): 18–23.

Barnard, Neil. "World Cancer Day: How Meat Can Be Murder." *The Independent*, February 4, 2013.

Benson, Janet E. "The Effects of Packinghouse Work on Southeast Asian Refugee Families." In *Newcomers in the Workplace: Immigrants and the Restructuring of the U.S. Economy*, ed. by Louise Lamphere, Alex Stepick, and Guillermo Grenier. Philadelphia: Temple University Press, 1994.

Berger, Peter L., and Thomas Luckmann. *The Social Construction of Reality: A Treatise in the Sociology of Knowledge*. New York: Doubleday, 1966.

Birke, Lynda. "Intimate Familiarities? Feminism and Human-Animal Studies." *Society & Animals* 10 (2002): 429–36.

Bjerklie, Steve. "Size Matters: The Meat Industry and the Corruption of Darwinian Economics." In *The CAFO Reader: The Tragedy of Industrial Animal Factories*, ed. by Daniel Imhoff. Healdsburg, CA: Watershed Media, 2010.

Black, R. "Stress and the Correctional Officer." *Police Stress* 5, no. 1 (1982): 10–16.

Boston Globe. "Animal Cruelty: Attacking the Messenger." April 15, 2013.

Brantz, Dorothee. "Animal Bodies, Human Health, and the Reform of Slaughterhouses in Nineteenth-Century Britain." In *Meat, Modernity, and the Rise of the Slaughterhouse*, ed. by Paula Young Lee. Lebanon: University of New Hampshire Press, 2008.

Bristow, Elizabeth, and Amy J. Fitzgerald. "Global Climate Change and the Industrial Animal Agriculture Link: The Construction of Risk." *Society & Animals* 19 (2011): 205–24.

Broadway, Michael J. "Bad to the Bone: The Social Costs of Beef Packing's Move to Rural Alberta." In *Writing Off the Rural West: Globalization, Governments, and the Transformation of Rural Communities*, ed. by Roger Epp and Dave Whitson. Edmonton: University of Alberta Press, 2001.

———. "Meatpacking and Its Social and Economic Consequences for Garden City, Kansas in the 1980s." *Urban Anthropology and Studies of Cultural Systems and World Economic Development* 19 (1990): 321–44.

———. "Meatpacking and the Transformation of Rural Communities: A Comparison of Brooks, Alberta and Garden City, Kansas." *Rural Sociology* 72 (2007): 560–82.

———. "Planning for Change in Small Towns or Trying to Avoid the Slaughterhouse Blues." *Journal of Rural Studies* 16 (2000): 37–46.

———. "What Happens When the Meatpackers Come to Town?" *Small Town* 24, no. 4 (1994): 24–28.

———. "Where's the Beef? The Integration of the Canadian and American Beefpacking Industries." *Prairie Forum* 23, no. 1 (1998): 19–30.

Broadway, Michael J., and Donald D. Stull. "'I'll Do Whatever You Want, but It Hurts': Worker Safety and Community Health in Modern Meatpacking." *Labor* 5, no. 2 (2008): 27–37.

———. "Meat Processing and Garden City, KS: Boom and Bust." *Journal of Rural Studies* 22 (2006): 55–66.

Brueggemann, John, and Cliff Brown. "The Decline of Industrial Unionism in the Meatpacking Industry: Event-Structure Analyses of Labor Unrest, 1946–1987." *Work and Occupations* 30 (2003): 327–60.

Bulliet, Richard W. *Hunters, Herders, and Hamburgers: The Past and Future of Human-Animal Relationships*. New York: Columbia University Press, 2005.

Burmeister, Larry L. "Lagoons, Litter and the Law: CAFO Regulations as Social Risk Politics." *Southern Rural Sociology* 18, no. 2 (2002): 56–87.

Campbell, T. Colin, and Thomas M. Campbell. *The China Study: The Most Comprehensive Study of Nutrition Ever Conducted and the Startling Implications for Diet, Weight Loss, and Long-term Health*. Dallas: BenBella Books, 2006.

Cartmill, Matt. *A View to a Death in the Morning: Hunting and Nature through History*. Cambridge, MA: Harvard University Press, 1993.

Cassuto, David N. "Bred Meat: The Cultural Foundation of the Factory Farm." *Law and Contemporary Problems* 70, no. 1 (2007): 59–87.

CBC News. "Seal Product Ban Upheld on 'Ethical' Grounds: 'Public Moral Concerns' Cited in Split Decision by the World Trade Organization." November 25, 2013. http://www.cbc.ca/news/canada/newfoundland-labrador/seal-product-ban-upheld-on-ethical-grounds-1.2438904.

CDC (Centers for Disease Control and Prevention). "CDC 2011 Estimates: Findings." January 8, 2014. http://www.cdc.gov/foodborneburden/2011-foodborne-estimates.html.

———. "Estimates of Foodborne Illness in the United States." January 8, 2014. http://www.cdc.gov/foodborneburden/.

———. "Prion Diseases." 2013. http://www.cdc.gov/ncidod/dvrd/prions/.

———. "Updated CDC Estimates of 2009 H1N1 Influenza Cases, Hospitalizations and Deaths in the United States, April 2009–April 10, 2010." http://www.cdc.gov/h1n1flu/pdf/CDC_2009_H1N1_Est_PDF_May_4_10_fulltext.pdf.

———. "Vital Signs: Incidence and Trends of Infection with Pathogens Transmitted Commonly Through Food—Foodborne Diseases Active Surveillance Network, 10 U.S. Sites, 1996–2010." June 10, 2011. http://www.cdc.gov/mmwr/preview/mmwrhtml/mm6022a5.htm?s_cid=mm6022a5_w.

Centner, Terence J. "Governmental Oversight of Discharges from Concentrated Animal Feeding Operations." *Environmental Management* 37, no. 6 (2006): 745–52.

Chambers, J. D., and G. E. Mingay. *The Agricultural Revolution, 1750–1880*. New York: Schocken Books, 1966.

Clark, Jonathan L. "Ecological Biopower, Environmental Violence against Animals, and the 'Greening' of the Factory Farm." *Journal for Critical Animal Studies* 10, no. 4 (2012): 109–29.

Clay, Jason W. *World Agriculture and the Environment: A Commodity-by-Commodity Guide to Impacts and Practices*. Washington, DC: Island Press, 2004.

Clutton-Brock, Juliet. *Animals as Domesticates: A World View through History*. East Lansing: Michigan State University Press, 2012.

Cochran, John K., and Mitchell B. Chamlin. "Deterrence and Brutalization: The Dual Effects of Executions." *Justice Quarterly* 17 (2000): 685–706.

Cochran, John K., Mitchell B. Chamlin, and Mark Seth. "Deterence or Brutalization? An Impact Assessment of Oklahoma's Return to Capital Punishment." *Criminology* 32 (1994): 107–34.

Cole, Matthew. "From 'Animal Machines' to 'Happy Meat'? Foucault's Ideas of Disciplinary and Pastoral Power Applied to 'Animal-Centered' Welfare Discourse." *Animals* 1 (2011): 83–101.

Cook, Christopher. "Sliced and Diced: The Labor You Eat." In *The CAFO Reader: The Tragedy of Industrial Animal Factories*, ed. by Daniel Imhoff. Healdsburg, CA: Watershed Media, 2010.

Cotton, William R., Jr. *Overshoot: The Ecological Basis of Revolutionary Change*. Urbana: University of Illinois Press, 1980.

Croney, Candace C. "Words Matter: Implications of Semantics and Imagery in Framing Animal-Welfare Issues." *Journal of Veterinary Medical Education* 37 (2010): 101–6.

Cronon, William. *Nature's Metropolis: Chicago and the Great West*. New York: W. W. Norton, 1991.

D'Silva, Joyce. "Adverse Impact of Industrial Animal Agriculture on the Health and Welfare of Farmed Animals." *Integrative Zoology* 1 (2006): 53–58.

Dalla, Rochelle L., Amy Ellis, and Sheran C. Cramer. "Immigration and Rural America: Latinos' Perceptions of Work and Residence in Three Meatpacking Communities." *Community, Work & Family* 8 (2005): 163–85.

Dauvergne, Peter. *The Shadows of Consumption: Consequences for the Global Environment*. Cambridge, MA: MIT Press, 2008.

Delgado, Christopher L. "Rising Consumption of Meat and Milk in Developing Countries Has Created a New Food Revolution." *Journal of Nutrition* 133 (2003): 3907S–3910S.

Delgado, Christopher L., and Clare A. Narrod. *Livestock Industrialization, Trade and Social-Health-Environment Impacts in Developing Countries*. June 28, 2002. http://www.fao.org/wairdocs/lead/x6115e/x6115e00.htm.

Dickes, Lori A., and Allen L. Dickes. "Oligopolists Then and Now: A Study of the Meatpacking Industry." *Journal of Economics and Economic Education Research* 4, no. 1 (2003): 95–110.

Dillard, Jennifer. "A Slaughterhouse Nightmare: Psychological Harm Suffered by Slaughterhouse Employees and the Possibility of Redress through Legal Reform." *Georgetown Journal on Poverty Law and Policy* 15 (2008): 391–409.

Donham, Kelley J., Steven Wing, David Osterberg, Jan L. Flora, Carole Hodne, Kendall M. Thu, and Peter S. Thorne. "Community Health and Socioeconomic Issues Surrounding Concentrated Animal Feeding Operations." *Environmental Health Perspectives* 115 (2007): 317–20.

Dunayer, Joan. *Animal Equality: Language and Liberation*. Derwood, MD: Ryce Publishing, 2001.

Earth Policy Institute. "Food and Agriculture." 2014. http://www.earth-policy.org/data_center/C24.

Edward, Bob, and Anthony E. Ladd. "Environmental Justice, Swine Production and Farm Loss in North Carolina." *Sociological Spectrum* 20 (2000): 263–90.

Eisnitz, Gail A. *Slaughterhouse: The Shocking Story of Greed, Neglect, and Inhumane Treatment inside the U.S. Meat Industry*. Amherst, NY: Prometheus Books, 1997.

Elias, Norbert. *The Civilizing Process: Sociogenetic and Psychogenetic Investigations*. 2nd ed. Oxford: Wiley-Blackwell, 2000.

Emhan, Abdurrahim, Ahmet Şener Yildiz, Yasin Bez, and Said Kingir. "Psychological Symptom Profile of Butchers Working in Slaughterhouse and Retail Meat Packing Business: A Comparative Study." *Kafkas Universitesi Veteriner Fakültesi Dergisi* 18 (2012): 319–22.

EPA (U.S. Environmental Protection Agency). "Regulatory Definitions of Large CAFOs, Medium CAFOs, and Small CAFOs." http://www.epa.gov/npdes/pubs/sector_table.pdf.

Environmental Working Group. "*Farm Subsidies*." The United States Summary Information." 2012. http://farm.ewg.org/region.php?fips=00000.

FAO (Food and Agriculture Organization of the United Nations). *The State of Food and Agriculture: Livestock in the Balance*. 2009. http://www.fao.org/docrep/012/i0680e/i0680e.pdf.

Feldstein, Stephanie. "Meet George Jetson—and His Test Tube Turkey?" *Huffington Post*, November 25, 2013. http://www.huffingtonpost.com/stephanie-feldstein/meet-george-jetson_b_4333534.html.

Festinger, Leon. *A Theory of Cognitive Dissonance*. Stanford: Stanford University Press, 1957.

Fink, Deborah. *Cutting into the Meatpacking Line: Workers and Change in the Rural Midwest*. Chapel Hill: University of North Carolina Press, 1998.

Fitzgerald, Amy J. "A Social History of the Slaughterhouse: From Inception to Contemporary Implications." *Human Ecology Review* 17, no. 1 (2010): 58–69.

———. "The 'Underdog' as 'Ideal Victim'? The Attribution of Victimhood in the 2007 Pet Food Recall." *International Review of Victimology* 17 (2010): 131–57.

Fitzgerald, Amy J., and Lori B. Baralt. "Media Constructions of Responsibility for the Production and Mitigation of Environmental Harms: The Case of Mercury-Contaminated Fish." *Canadian Journal of Criminology and Criminal Justice* 52 (2010): 341–68.

Fitzgerald, Amy J., Linda Kalof, and Thomas Dietz. "Slaughterhouses and Increased Crime Rates: An Empirical Analysis of the Spillover from 'The Jungle' into the Surrounding Community." *Organization and Environment* 22 (2009): 158–84.

Fitzgerald, Amy J., and Nik Taylor. "The Cultural Hegemony of Meat and the Animal Industrial Complex." In *The Rise of Critical Animal Studies: From the Margins to the Centre*, ed. by Nik Taylor and Richard Twine. London: Palgrave Macmillan, 2014.

Fitzgerald, Deborah Kay. *Every Farm a Factory: The Industrial Ideal in American Agriculture*. Yale Agrarian Studies Series. New Haven: Yale University Press, 2003.

Foer, Jonathan Safran. *Eating Animals*. New York: Little, Brown and Company, 2009.

Foltz, Richard C. *Animals in Islamic Tradition and Muslim Cultures*. Oxford: Oneworld Publications, 2006.

Francione, Gary L. "Animals, Property and Legal Welfarism: 'Unnecessary' Suffering and the 'Humane' Treatment of Animals." *Rutgers Law Review* 46 (1994): 721–70.

———. "Animals, Property, and Personhood." In *People, Property, or Pets?*, ed. by Marc D. Hauser, Fiery Cushman, and Matthew Kamen. West Lafayette: Purdue University Press, 2006.

———. *Introduction to Animal Rights: Your Child or the Dog?* Philadelphia: Temple University Press, 2000.

———. *Rain without Thunder: The Ideology of the Animal Rights Movement.* Philadelphia: Temple University Press, 1996.

Franklin, Adrian. *Animals and Modern Cultures: A Sociology of Human-Animal Relations in Modernity.* Thousand Oaks: Sage, 1999.

Gaard, Greta Claire. "Vegetarian Ecofeminism: A Review Essay." *Frontiers* 23, no. 3 (2002): 117–46.

Gilchrist, Mary J., Christina Greko, David B. Wallinga, George W. Beran, David G. Riley, and Peter S. Thorne. "The Potential Role of Concentrated Animal Feeding Operations in Infectious Disease Epidemics and Antibiotic Resistance." *Environmental Health Perspectives* 115 (2007): 313–16.

Glenn, Cathy B. "Constructing Consumables and Consent: A Critical Analysis of Factory Farm Industry Discourse." *Journal of Communication Inquiry* 28 (2004): 63–81.

Goldschmidt, Walter. *As You Sow: Three Studies in the Social Consequences of Agribusiness.* Montclair, NJ: Allanheld, Osmun, 1978.

Gossard, Marcia Hill, and Richard York. "Social Structural Influences on Meat Consumption." *Human Ecology Review* 10, no. 1 (2003): 1–9.

Gouveia, Lourdes, and Arunas Juska. "Taming Nature, Taming Workers: Constructing the Separation between Meat Consumption and Meat Production in the U.S." *Sociologia Ruralis* 42 (2002): 370–90.

Green, Christopher F., Shawn G. Gibbs, Patrick M. Tarwater, Linda C. Mota, and Pasquale V. Scarpino. "Bacterial Plume Emanating from the Air Surrounding Swine Confinement Operations." *Journal of Occupational and Environmental Hygiene* 3 (2006): 9–15.

Greene, Catherine, Carolyn Dimitri, Biing-Hwan Lin, William McBride, Lydia Oberholtzer, and Travis Smith. "Emerging Issues in the US Organic Industry." US Department of Agriculture, Economic Research Service. June 2009. http://www.ers.usda.gov/media/155923/eib55_1_.pdf.

Gregory, Neville G. *Animal Welfare and Meat Production.* 2nd ed. Cambridge, MA: CABI, 2007.

Grey, Mark. "Pork, Poultry and Newcomers in Storm Lake, Iowa." In *Any Way You Cut It: Meat Processing and Small-Town America*, ed. by Donald D. Stull, Michael J. Broadway, and David Griffith. Lawrence: University Press of Kansas, 1995.

———. "Meatpacking in Storm Lake, Iowa: A Community in Transition." In *Pigs, Profits, and Rural Communities*, ed. by Kendall M. Thu and E. Paul Durrenberger. Albany: State University of New York Press, 1998.

Gunderson, Ryan. "From Cattle to Capital: Exchange Value, Animal Commodification, and Barbarism." *Critical Sociology* 39 (2013): 259–75.

———. "The Metabolic Rifts of Livestock Agribusiness." *Organization and Environment* 24 (2011): 404–22.

Hahn Niman, Nicolette. *Righteous Porkchop: Finding a Life and Good Food beyond Factory Farms.* New York: HarperCollins, 2009.

Halpern, Rick. *Down on the Killing Floor: Black and White Workers in Chicago's Packinghouses, 1904–54.* Urbana: University of Illinois Press, 1997.

Hardeman, Egbert, and Henk Jochemsen. "Are There Ideological Aspects to the Modernization of Agriculture?" *Journal of Agricultural and Environmental Ethics* 25 (2012): 657–74.

Harrison, Ruth. *Animal Machines: The New Factory Farming Industry.* New York: Ballantine Books, 1964.

Headlee, Sue. *The Political Economy of the Family Farm: The Agrarian Roots of American Capitalism.* New York: Praeger, 1991.

Heinz, Bettina, and Ronald Lee. "Getting Down to the Meat: The Symbolic Construction of Meat Consumption." *Communication Studies* 49 (1998): 86–99.

Hemsworth, Paul H., and Grahame J. Coleman. *Human-Livestock Interactions: The Stockperson and the Productivity and Welfare of Intensively Farmed Animals.* Cambridge, MA: CABI, 2011.

Hennessy, David A. "Slaughterhouse Rules: Animal Uniformity and Regulating for Food Safety in Meat Packing." *American Journal of Agricultural Economics* 87 (2005): 600–609.

Herzog, Hal. *Some We Love, Some We Hate, Some We Eat: Why It's So Hard to Think Straight about Animals.* New York: Harper Perennial, 2011.

Herzog, Harold A., and Sandy McGee. "Psychological Aspects of Slaughter: Reactions of College Students to Killing and Butchering Cattle and Hogs." *International Journal for the Study of Animal Problems* 4, no. 2 (1983): 124–32.

HHS (U.S. Department of Health and Human Services). "2012 HHS Poverty Guidelines." 2012. http://aspe.hhs.gov/poverty/12poverty.shtml.

Horowitz, Roger. *Putting Meat on the American Table: Taste, Technology, Transformation.* Baltimore: Johns Hopkins University Press, 2006.

———. "'Where Men Will Not Work': Gender, Power, Space, and the Sexual Division of Labor in America's Meatpacking Industry, 1890–1990." *Technology and Culture* 38 (1997): 187–213.

Horowitz, Roger, and Mark J. Miller. *Immigrants in the Delmarva Poultry Processing Industry: The Changing Face of Georgetown, Delaware, and Environs.* JSRI Occasional Paper No. 37. East Lansing: Julian Samora Research Institute, Michigan State University, 1999.

Horrigan, Leo, Robert S. Lawrence, and Polly Walker. "How Sustainable Agriculture Can Address the Environmental and Health Harms of Industrial Agriculture." *Environmental Health Perspectives* 110 (2002): 445–56.

Hughes, Everett C. "Good People and Dirty Work." *Social Problems* 10 (1962): 3–11.

Ibrahim, Darian M. "A Return to Descartes: Property, Profit, and the Corporate Ownership of Animals." *Law and Contemporary Problems* 70, no. 1 (2007): 89–115.

Imhoff, Daniel, ed. *The CAFO Reader: The Tragedy of Industrial Animal Factories.* Healdsburg, CA: Watershed Media, 2010.

———. "Introduction." In *The CAFO Reader: The Tragedy of Industrial Animal Factories*, ed. by Daniel Imhoff. Healdsburg, CA: Watershed Media, 2010.

Isaacs-Blundin, Cecilia. "Why Manure May Be the Farm Animal Advocate's Best Friend: Using Environmental Statutes to Access Factory Farms." *Journal of Animal Law and Ethics* 2 (2007): 173–87.

Jablonsky, Thomas J. *Pride in the Jungle: Community and Everyday Life in Back of the Yards Chicago.* Baltimore: Johns Hopkins University Press, 1993.

Jasper, James M., and Dorothy Nelkin. *The Animal Rights Crusade: The Growth of a Moral Protest.* New York: Free Press, 1992.

Johnsen, Carolyn. *Raising a Stink: The Struggle over Factory Hog Farms in Nebraska.* Lincoln: University of Nebraska Press, 2003.

Joy, Melanie. *Why We Love Dogs, Eat Pigs and Wear Cows: An Introduction to Carnism.* San Francisco: Conari Press, 2010.

Juska, Arunas, Lourdes Gouveia, Jackie Gabriel, and Kathleen P. Stanley. "Manufacturing Bacteriological Contamination Outbreaks in Industrialized Meat Production Systems: The Case of *E. coli* O157:H7." *Agriculture and Human Values* 20 (2003): 3–19.

Kalechofsky, Roberta, ed. *Judaism and Animal Rights: Classical and Contemporary Responses*. Marblehead, MA: Micah Publications, 1992.

Kalof, Linda. *Looking at Animals in Human History*. London: Reaktion, 2007.

Kalof, Linda, Thomas Dietz, Paul C. Stern, and Gregory A. Guagnano. "Social Psychological and Structural Influences on Vegetarian Beliefs." *Rural Sociology* 64 (1999): 500–511.

Kauffman, Kelsey. *Prison Officers and Their World*. Cambridge, MA: Harvard University Press, 1988.

Kay, Jared D. "Butchers, Tanners, and Tallow Chandlers: The Geography of Slaughtering in Early Nineteenth-Century New York City." In *Meat, Modernity, and the Rise of the Slaughterhouse*, ed. by Paula Young Lee. Lebanon: University of New Hampshire Press, 2008.

Kendall, Holli A., Linda A. Lobao, and Jeff S. Sharp. "Public Concern with Animal Well-Being: Place, Social Structural Location, and Individual Experience." *Rural Sociology* 71 (2006): 399–428.

Kennedy, Robert F., Jr. "Foreword." In Nicolette Hahn Niman, *Righteous Porkchop: Finding a Life and Good Food beyond Factory Farms*. New York: HarperCollins, 2009.

Kilpatrick, John A. "Concentrated Animal Feeding Operations and Proximate Property Values." *Appraisal Journal* 69 (2001): 301–6.

Kimbrell, Andrew. "Cold Evil: The Ideologies of Industrialism." In *The CAFO Reader: The Tragedy of Industrial Animal Factories*, ed. by Daniel Imhoff. Healdsburg, CA: Watershed Media, 2010.

King, David R. "The Brutalization Effect: Execution Publicity and the Incidence of Homicide in South Carolina." *Social Forces* 57 (1978): 683–87.

Kirkhorne, Steven, and Marc B. Schenker. "Human Health Effects of Agriculture: Physical Diseases and Illnesses." National Ag Safety Database, 2001. http://nasdonline.org/document/1836/d001772/human-health-effects-of-agriculture-physical-diseases-and.html.

Ladd, Anthony E., and Bob Edward. "Corporate Swine and Capitalist Pigs: A Decade of Environmental Injustice and Protest in North Carolina." *Social Justice* 29, no. 3 (2002): 26–46.

Lappé, Anna. "Diet for a Hot Planet: Livestock and Climate Change." In *The CAFO Reader: The Tragedy of Industrial Animal Factories*, ed. by Daniel Imhoff. Healdsburg, CA: Watershed Media, 2010.

Longo, Stefano B., and Nicholas Malone. "Meat, Medicine, and Materialism: A Dialectical Analysis of Human Relationships to Nonhuman Animals and Nature." *Human Ecology Review* 13, no. 2 (2006): 111–21.

Lorber, Judith. *Paradoxes of Gender*. New Haven: Yale University Press, 1994.

MacLachlan, Ian. "Humanitarian Reform, Slaughter Technology, and Butcher Resistance in Nineteenth-Century Britain." In *Meat, Modernity, and the Rise of the Slaughterhouse*, ed. by Paula Young Lee. Lebanon: University of New Hampshire Press, 2008.

MacNair, Rachel M. *Perpetration-Induced Traumatic Stress: The Psychological Consequences of Killing*. Westport: Praeger, 2002.

Magdoff, Fred, John Bellamy Foster, and Frederick H. Buttel, eds. *Hungry for Profit: The Agribusiness Threat to Farmers, Food, and the Environment*. New York: Monthly Review Press, 2000.

Mallon, Robyn. "The Deplorable Standard of Living Faced by Farmed Animals in America's Meat Industry and How to Improve Conditions by Eliminating the Corporate Farm." *MSU Journal of Medicine and Law* 9 (2005): 389–415.

Marcus, Erik. *Meat Market: Animals, Ethics, and Money*. Boston: Brio Press, 2005.

Marshall, Amy D., Jillian Panuzio, and Casey T. Taft. "Intimate Partner Violence among Military Veterans and Active Duty Servicemen." *Clinical Psychology Review* 25 (2005): 862–76.

Marshall, David, and Marilyn McShane. "First to Fight: Domestic Violence and the Subculture of the Marine Corps." In *Battle Cries on the Home Front: Violence in the Military Family*, ed. by Peter J. Mercier and Judith D. Mercier. Springfield, IL: Charles C. Thomas, 2000.

Mason, Jim, and Mary Finelli. "Brave New Farm?" In *In Defense of Animals: The Second Wave*, ed. by Peter Singer. Oxford: Wiley-Blackwell, 2005.

Matheny, Gaverick, and Cheryl Leahy. "Farm-Animal Welfare, Legislation, and Trade." *Law and Contemporary Problems* 70, no. 1 (2007): 325–58.

Maurer, Donna. *Vegetarianism: Movement or Moment?* Philadelphia: Temple University Press, 2002.

Mayfield, L. E., R. M. Bennett, R. B. Tranter, and M. J. Wooldridge. "Consumption of Welfare-Friendly Food Products in Great Britain, Italy and Sweden, and How It May Be Influenced by Consumer Attitudes to, and Behaviour towards, Animal Welfare Attributes." *International Journal of Sociology of Agriculture and Food* 15 (2007): 59–73.

Mazoyer, Marcel, and Laurence Roudart. *A History of World Agriculture: From the Neolithic Age to the Current Crisis*. London: Earthscan, 2006.

McBride, William D., and Nigel Key. "U.S. Hog Production from 1992 to 2009: Technology, Restructuring, and Productivity Growth." U.S. Department of Agriculture, Economic Research Service. October 2013. http://www.ers.usda.gov/media/1207987/err158.pdf.

McNeil, Ian, ed. *An Encyclopedia of the History of Technology*. New York: Routledge, 1996.

Meens, Rob. "Eating Animals in the Early Middle Ages: Classifying the Animal World and Building Group Identities." In *The Animal/Human Boundary: Historical Perspectives*, ed. by Angela N. H. Creager and William Chester Jordan. Woodbridge: University of Rochester Press, 2002.

Mench, Joy A., Harvey James, Edmond J. Pajor, and Paul B. Thompson. "The Welfare of Animals in Concentrated Animal Feeding Operations." Food and Agriculture Organization of the United Nations. August 30, 2010. http://www.fao.org/ag/againfo/themes/animal-welfare/aw-awhome/detail/pt/item/11934/icode/en/.

Mercier, Peter J. "Violence in the Military Family." In *Battle Cries on the Home Front: Violence in the Military Family*, ed. by Peter J. Mercier and Judith D. Mercier. Springfield, IL: Charles C. Thomas, 2000.

Mertig, Angela G., and Riley E. Dunlap. "Environmentalism, New Social Movements, and the New Class: A Cross-National Investigation." *Rural Sociology* 66 (2001): 113–36.

Mirabelli, Maria C., Steve Wing, Stephen W. Marshall, and Timothy C. Wilcosky. "Asthma Symptoms among Adolescents Who Attend Public Schools That Are Located Near Confined Swine Feeding Operations." *Pediatrics* 118 (2006): 66–75.

Mithen, Steven. "The Hunter-Gatherer Prehistory of Human-Animal Interactions." *Anthrozoos* 12 (1999): 195–204.

National Farmers Union News, "Farmers Union Members Testify on Negative Impacts of Livestock Concentration." 49, no. 8 (2002): 3.

National Hog Farmer. "Consumer Survey Outlines Pork Perception Issues." December 15, 2012. http://nationalhogfarmer.com/animal-well-being/consumer-survey-outlines-pork-perception-issues.

National Labor Relations Board. "National Labor Relations Act." http://www.nlrb.gov/national-labor-relations-act.

New York Times. "Eating with Our Eyes Closed." April 9, 2013.

———. "What Meat Means." February 6, 2005.

Nibert, David. *Animal Oppression and Human Violence: Domesecration, Capitalism, and Global Conflict*. New York: Columbia University Press, 2013.

———. *Animal Rights/Human Rights: Entanglements of Oppression and Liberation*. Lanham, MD: Rowman and Littlefield Publishers, 2002.

Nierenberg, Danielle. "Factory Farming in the Developing World." *World-Watch*. May/June 2003. http://www.worldwatch.org/system/files/EP163A.pdf.

Noble, Martha. "Paying the Polluters: Animal Factories Feast on Taxpayer Subsidies." In *The CAFO Reader: The Tragedy of Industrial Animal Factories*, ed. by Daniel Imhoff. Healdsburg, CA: Watershed Media, 2010.

Noske, Barbara. *Humans and Other Animals: Beyond the Boundaries of Anthropology*. London: Pluto Press, 1989.

Novek, Joel. "Intensive Hog Farming in Manitoba: Transnational Treadmills and Local Conflicts." *Canadian Review of Sociology* 40 (2003): 3–26.

Olsson, Karen. "The Shame of Meatpacking." *The Nation* 275, no. 8 (2002): 11–16.

Otter, Chris. "Civilizing Slaughter: The Development of the British Public Abattoir, 1850–1910." In *Meat, Modernity, and the Rise of the Slaughterhouse*, ed. by Paula Young Lee. Lebanon: University of New Hampshire Press, 2008.

Pachirat, Timothy. *Every Twelve Seconds: Industrialized Slaughter and the Politics of Sight*. New Haven: Yale University Press, 2011.

Packwood Freeman, Carrie. "Meat's Place on the Campaign Menu: How US Environmental Discourse Negotiates Vegetarianism." *Environmental Communication* 4 (2010): 255–76.

Pacyga, Dominic A. "Chicago: Slaughterhouse to the World." In *Meat, Modernity, and the Rise of the Slaughterhouse*, ed. by Paula Young Lee. Lebanon: University of New Hampshire Press, 2008.

Parry, Jovian. "Gender and Slaughter in Popular Gastronomy." *Feminism and Psychology* 20 (2010): 381–96.

Patterson, Charles. *Eternal Treblinka: Our Treatment of Animals and the Holocaust*. New York: Lantern Books, 2002.

Pearce, R. A., J. J. Sheridan, and D. J. Bolton. "Distribution of Airborne Microorganisms in Commercial Pork Slaughter Processes." *International Journal of Food Microbiology* 107 (2006): 186–91.

Perkel, Colin. "University of Guelph 'Enviropigs' Put Down, Critics Blast 'Callous' Killing." *Huffington Post, Canada*, June 21, 2012. http://www.huffingtonpost.ca/2012/06/21/enviropigs-university-of -guelph_n_1617140.html.

Phelps, Norm. *The Great Compassion: Buddhism and Animal Rights*. New York: Lantern Books, 2004.

Philo, Chris. "Animals, Geography, and the City: Notes on Inclusions and Exclusions." In *Animal Geographies: Place, Politics, and Identity in the Nature-Culture Borderlands*, ed. by Jennifer Wolch and Jody Emel. New York: Verso, 1998.

Pollan, Michael. *The Omnivore's Dilemma: A Natural History of Four Meals*. New York: Penguin Books, 2006.

———. "Power Steer: On the Trail of Industrial Beef." In *The CAFO Reader: The Tragedy of Industrial Animal Factories*, ed. by Daniel Imhoff. Healdsburg, CA: Watershed Media, 2010.

Ponette-González, Alexandra G., and Matthew Fry. "Pig Pandemic: Industrial Hog Farming in Eastern Mexico." *Land Use Policy* 27 (2010): 1107–10.

Price, T. Douglas, and Anne Birgitte Gebauer. "New Perspectives on the Transition to Agriculture." In *Last Hunters, First Farmers: New Perspectives on the Prehistoric Transition to Agriculture*, ed. by T. Douglas Price and Anne Birgitte Gebauer. Santa Fe: School of American Research Press, 1996.

Purdy, Jedediah. "Open the Slaughterhouses." *New York Times*, April 8, 2013.

Ransom, Elizabeth. "The Rise of Agricultural Animal Welfare Standards as Understood through a Neo-Institutional Lens." *International Journal of Sociology of Agriculture and Food* 15 (2007): 26–44.

Regan, Tom. "Preface." In Sue Coe, *Dead Meat*. New York: Four Walls Eight Windows, 1996.

Rémy, Catherine. "Une mise à mort industrielle 'humaine'? L'abattoir ou l'impossible objectivation des animaux." *Politix* 16, no. 64 (2003): 51–73.

Rifkin, Jeremy. *Beyond Beef: The Rise and Fall of the Cattle Culture*. New York: Penguin Books, 1992.

Rogers, Richard A. "Beasts, Burgers, and Hummers: Meat and the Crisis of Masculinity in Contemporary Television Advertisements." *Environmental Communication* 2 (2008): 281–301.

Rollin, Bernard. "Farm Factories: The End of Animal Husbandry." In *The CAFO Reader: The Tragedy of Industrial Animal Factories*, ed. by Daniel Imhoff. Healdsburg, CA: Watershed Media, 2010.

Rosen, Leora N., Robert J. Kaminski, Angela Moore Parmley, Kathryn H. Knudson, and Peggy Fancher. "The Effects of Peer Group Climate on Intimate Partner Violence among Married Male U.S. Army Soldiers." *Violence Against Women* 9 (2003): 1045–71.

Rosen, Steven J. *Holy Cow: The Hare Krishna Contribution to Vegetarianism and Animal Rights*. New York: Lantern Books, 2004.

Rothgerber, Hank. "Real Men Don't Eat (Vegetable) Quiche: Masculinity and the Justification of Meat Consumption." *Psychology of Men and Masculinities* 14, no. 4 (2013): 363–75.

Ruby, Matthew B. "Vegetarianism: A Blossoming Field of Study." *Appetite* 58 (2012): 141–50.

Sabin, Louis. *Agriculture*. New York: Troll Communications, 1984.

Saenz, Roberto A., Herbert W. Hethcote, and Gregory C. Gray. "Confined Animal Feeding Operations as Amplifiers of Influenza." *Vector-Borne and Zoonotic Diseases* 6 (2006): 338–46.

Sapontzis, Steve F., ed. *Food for Thought: The Debate over Eating Meat*. Amherst: Prometheus Books, 2004.

Schlosser, Eric. "Bad Meat: Deregulation Makes Eating a High-Risk Behavior." In *The CAFO Reader: The Tragedy of Industrial Animal Factories*, ed. by Daniel Imhoff. Healdsburg, CA: Watershed Media, 2010.

———. *Fast Food Nation: The Dark Side of the All-American Meal*. New York: Houghton Mifflin, 2005.

Schmidt, Charles W. "Low on the Hog: The Quality of Life near Swine Farms." *Environmental Health Perspectives* 108 (2000): A134.

———. "Swine CAFOs and Novel H1N1 Flu: Separating Facts from Fears." *Environmental Health Perspectives* 117 (2009): A394–A401.

Schnaiberg, Allan. *The Environment: From Surplus to Scarcity*. New York: Oxford University Press, 1980.

Scully, Matthew. *Dominion: The Power of Man, the Suffering of Animals, and the Call to Mercy*. New York: St. Martin's Press, 2002.

———. "Fear Factories: The Case for Compassionate Conservatism—for Animals." In *The CAFO Reader: The Tragedy of Industrial Animal Factories*, ed. by Daniel Imhoff. Healdsburg, CA: Watershed Media, 2010.

Serpell, James. *In the Company of Animals: A Study of Human-Animal Relationships*. Oxford: Basil Blackwell, 1986.

———. "Having Our Dogs and Eating Them Too: Why Animals Are a Social Issue." *Journal of Social Issues* 65 (2009): 633–44.

Shukin, Nicole. *Animal Capital: Rendering Life in Biopolitical Times*. Minneapolis: University of Minnesota Press, 2009.

Sinclair, Upton. *The Jungle*. 4th ed. New York: Viking Press, 1946.

Singer, Peter, and Jim Mason. *The Way We Eat: Why Our Food Choices Matter*. Emmaus, PA: Rodale, 2006.

Skaggs, Jimmy M. *Prime Cut: Livestock Raising and Meatpacking in the United States, 1607–1983*. College Station: Texas A & M University Press, 1986.

Smith, Mick. "The 'Ethical' Space of the Abattoir: On the (In)human(e) Slaughter of Other Animals." *Human Ecology Review* 9, no. 2 (2002): 49–58.

Snider, Laureen. "Cooperative Models and Corporate Crime: Panacea or Cop-out?" *Crime and Delinquency* 36 (1990): 373–90.

Sorenson, John. *About Canada: Animal Rights*. Halifax: Fernwood Publishing, 2010.

Stack, Steven J., and Olga Tsoudis. "Suicide Risk among Correctional Officers: A Logistic Regression Analysis." *Archives of Suicide Research* 3 (1997): 183–86.

Stănescu, Vasile. "'Green' Eggs and Ham? The Myth of Sustainable Meat and the Danger of the Local." In *Critical Theory and Animal Liberation*, ed. by John Sanbonmatsu. Toronto: Rowman and Littlefield, 2011.

Starmer, Elanor, Aimee Witteman, and Timothy A. Wise. "Feeding the Factory Farm: Implicit Subsidies to the Broiler Chicken Industry." Global Development and Environment Institute, Tufts University. June 2006. http://www.ase.tufts.edu/gdae/Pubs/wp/06-03BroilerGains.pdf.

Stathopoulos, Anastasia S. "You Are What Your Food Eats: How Regulation of Factory Farm Conditions Could Improve Human Health and Animal Welfare Alike." *New York University Journal of Legislation and Public Policy* 13 (2010): 407–44.

Steinfeld, Henning, Pierre Gerber, Tom Wassenaar, Vincent Castel, Mauricio Rosales, and Cees de Haan. "Livestock's Long Shadow: Environmental Issues and Options." Food and Agriculture Organization of the United Nations. 2006. http://www.fao.org/3/a-a0701e.pdf.

Stith, Pat, Joby Warrick, and Melanie Sill. "Boss Hog: The Power of Pork." *1996 Public Service: The News & Observer*. The Pulitzer Prize Organization. February 19, 1995. http://www.pulitzer.org/archives/5892.

Striffler, Steve. *Chicken: The Dangerous Transformation of America's Favorite Food*. Yale Agrarian Studies Series. New Haven: Yale University Press, 2005.

———. "Watching the Chickens Pass By: The Grueling Monotony of the Disassembly Line." In *The CAFO Reader: The Tragedy of Industrial Animal Factories*, ed. by Daniel Imhoff. Healdsburg, CA: Watershed Media, 2010.

Stuart, Diana, and Michelle R. Worosz. "The Myth of Efficiency: Technology and Ethics in Industrial Food Production." *Journal of Agricultural and Environmental Ethics* 26 (2013): 231–56.

———. "Risk, Anti-reflexivity, and Ethical Neutralization in Industrial Food Processing." *Agriculture and Human Values* 29 (2012): 287–301.

Stull, Donald D. "Knock 'Em Dead: Work on the Killfloor of a Modern Beefpacking Plant." In *Newcomers in the Workplace: Immigrants and the Restructuring of the U.S. Economy*, ed. by Louise Lamphere, Alex Stepick, and Guillermo Grenier. Philadelphia: Temple University Press, 1994.

Stull, Donald D., and Michael J. Broadway. "The Effects of Restructuring on Beefpacking in Kansas." *Kansas Business Review* 14, no. 1 (1990): 10–16.

———. *Slaughterhouse Blues: The Meat and Poultry Industry in North America*. Belmont, CA: Wadsworth, 2004.

Szasz, Andrew. *Shopping Our Way to Safety: How We Changed from Protecting the Environment to Protecting Ourselves*. Minneapolis: University of Minnesota Press, 2007.

Taquino, Michael, Domenico Parisi, and Duane A. Gill. "Units of Analysis and the Environmental Justice Hypothesis: The Case of Industrial Hog Farms." *Social Science Quarterly* 83 (2002): 298–316.

Thomas, Keith. *Man and the Natural World: A History of the Modern Sensibility*. New York: Pantheon Books, 1983.

Thompson, F. M. L. "The Second Agricultural Revolution, 1815–1880." *Economic History Review* 21 (1968): 62–77.

Thu, Kendall. "CAFOs Are in Everyone's Backyard: Industrial Agriculture, Democracy, and the Future." In *The CAFO Reader: The Tragedy of Industrial Animal Factories*, ed. by Daniel Imhoff. Healdsburg, CA: Watershed Media, 2010.

Tietz, Jeff. "Boss Hog: The Rapid Rise of Industrial Swine." In *The CAFO Reader: The Tragedy of Industrial Animal Factories*, ed. by Daniel Imhoff. Healdsburg, CA: Watershed Media, 2010.

Twine, Richard. *Animals as Biotechnology: Ethics, Sustainability and Critical Animal Studies*. Washington, DC: Earthscan, 2010.

———. "Revealing the 'Animal Industrial Complex'—A Concept and Method for Critical Animal Studies?" *Journal for Critical Animal Studies* 10, no. 1 (2012): 12–39.

———. "Searching for the 'Win-Win'? Animals, Genomics and Welfare." *International Journal of Sociology of Agriculture and Food* 15 (2007): 8–25.

USDA (U.S. Department of Agriculture). "Humane Methods of Slaughter Act." http://awic.nal.usda .gov/government-and-professional-resources/federal-laws/humane-methods-slaughter-act.

USDA (U.S. Department of Agriculture), Economic Research Service. "Livestock and Meat Domestic Data." 2013. http://www.ers.usda.gov/data-products/livestock-meat-domestic-data.

USDA (U.S. Department of Agriculture) and HHS (U.S. Department of Health and Human Services). "Dietary Guidelines for Americans 2010." Washington, DC: US Government Printing Office, 2010. http://www.health.gov/dietaryguidelines/dga2010/DietaryGuidelines2010.pdf.

U.S. Department of Labor, Bureau of Labor Statistics. "Labor Force Statistics from the Current Population Survey." February 26, 2014. http://www.bls.gov/cps/cpsaat11.htm.

———. "Occupational Employment and Wages—May 2013." April 1, 2014. http://www.bls.gov/ news.release/pdf/ocwage.pdf.

———. "Table SNR05: Incidence Rate and Number of Nonfatal Occupational Injuries by Industry and Ownership, 2011." http://www.bls.gov/iif/oshwc/osh/os/ostb3183.pdf.

U.S. General Accounting Office. "Animal Agriculture: Waste Management Practices." July 1999. http://www.gao.gov/archive/1999/rc99205.pdf.

U.S. Government Accountability Office. "Workplace Safety and Health: Safety in the Meat and Poultry Industry, while Improving, Could Be Further Strengthened." January 2005. http://www.gao .gov/new.items/d0596.pdf.

U.S. Poultry and Egg Association. "Economic Data." http://uspoultry.org/economic_data/.

University of Guelph. "Enviropig." 2010. http://www.uoguelph.ca/enviropig/.

Vanhonacker, Filiep, Wim Verbeke, Els Van Poucke, and Frank A. M. Tuyttens. "Segmentation Based on Consumers' Perceived Importance and Attitude toward Farm Animal Welfare." *International Journal of Sociology of Agriculture and Food* 15 (2007): 91–107.

Vialles, Noëlie. *Animal to Edible*, trans. by J. A. Underwood. Cambridge: Cambridge University Press, 1994.

Vining, Joseph. "Animal Cruelty Laws and Factory Farming." *Michigan Law Review First Impressions* 106, no. 5 (2008): 123–27.

Wallerstein, Immanuel. *The Modern World-System I: Capitalist Agriculture and the Origins of the European World-Economy in the Sixteenth Century*. Berkeley: University of California Press, 2011.

Warrick, Joby. "'They Die Piece by Piece': In Overtaxed Plants, Humane Treatment of Cattle Is Often a Battle Lost." *Washington Post*, April 10, 2001.

Weis, Tony. "The Accelerating Biophysical Contradictions of Industrial Capitalist Agriculture." *Journal of Agrarian Change* 10 (2010): 315–41.

West, Bridgett M., Peggy Liggit, Daniel L. Clemens, and Steven N. Francoeur. "Antibiotic Resistance, Gene Transfer, and Water Quality Patterns Observed in Waterways near CAFO Farms and Wastewater Treatment Facilities." *Water, Air and Soil Pollution* 217 (2011): 473–89.

West, Candace, and Sarah Fenstermaker. "Power, Inequality, and the Accomplishment of Gender: An Ethnomethodological View." In *Theory on Gender, Feminism on Theory*, ed. by Paula England. Hawthorne, NY: Aldine De Gruyter, 1993.

Wilkie, Rhoda. "Sentient Commodities and Productive Paradoxes: The Ambiguous Nature of Human–Livestock Relations in Northeast Scotland." *Journal of Rural Studies* 21 (2005): 213–30.

———. *Livestock/Deadstock: Working with Farm Animals from Birth to Slaughter.* Philadelphia: Temple University Press, 2010.

Williams, Nancy M. "Affected Ignorance and Animal Suffering: Why Our Failure to Debate Factory Farming Puts Us at Moral Risk." *Journal of Agricultural and Environmental Ethics* 21 (2008): 371–84.

Winders, Bill, and David Nibert. "Consuming the Surplus: Expanding 'Meat' Consumption and Animal Oppression." *International Journal of Sociology and Social Policy* 24, no. 9 (2004): 76–96.

Wing, Steve, Rachel Avery Horton, Naeema Muhammad, Gary R. Grant, Mansoureh Tajik, and Kendall M. Thu. "Integrating Epidemiology, Education, and Organizing for Environmental Justice: Community Health Effects of Industrial Hog Operations." *American Journal of Public Health* 98 (2008): 1390–97.

Wood, Ellen Meiksins. *The Origin of Capitalism: A Longer View.* London: Verso, 2002.

Woods, Michael, and Mary B. Woods. *Ancient Agriculture: From Foraging to Farming.* Minneapolis: Lerner Publications, 2000.

World Organisation for Animal Health. "OIE's Achievements in Animal Welfare." September 10, 2014. http://www.oie.int/animal-welfare/animal-welfare-key-themes.

Yang, Bijou, and David Lester. "The Deterrent Effect of Executions: A Meta-Analysis Thirty Years after Ehrlich." *Journal of Criminal Justice* 36 (2008): 453–60.

York, Richard. "Humanity and Inhumanity: Toward a Sociology of the Slaughterhouse." *Organization and Environment* 17 (2004): 260–65.

Young Lee, Paula. "Siting the Slaughterhouse: From Shed to Factory." In *Meat, Modernity, and the Rise of the Slaughterhouse*, ed. by Paula Young Lee. Lebanon: University of New Hampshire Press, 2008.

Zande, Karly. "Raising a Stink: Why Michigan CAFO Regulations Fail to Protect the State's Air and Great Lakes and Are in Need of Revision." *Buffalo Environmental Law Journal* 16, no. 1/2 (2008–9): 1–53.

Zogby, John, John Bruce, and Rebecca Wittman. "Nationwide Views on the Treatment of Farm Animals." Zogby International. October 22, 2003. http://civileats.com/wp-content/uploads/2009/09/AWT-final-poll-report-10-22.pdf.

Zuzworsky, Rose. "From the Marketplace to the Dinner Plate: The Economy, Theology, and Factory Farming." *Journal of Business Ethics* 29 (2001): 177–88.

Index